Science and Fiction

Science and Fiction – A Springer Series

This collection of entertaining and thought-provoking books will appeal equally to science buffs, scientists and science-fiction fans. It was born out of the recognition that scientific discovery and the creation of plausible fictional scenarios are often two sides of the same coin. Each relies on an understanding of the way the world works, coupled with the imaginative ability to invent new or alternative explanations—and even other worlds. Authored by practicing scientists as well as writers of hard science fiction, these books explore and exploit the borderlands between accepted science and its fictional counterpart. Uncovering mutual influences, promoting fruitful interaction, narrating and analyzing fictional scenarios, together they serve as a reaction vessel for inspired new ideas in science, technology, and beyond.

Whether fiction, fact, or forever undecidable: the Springer Series "Science and Fiction" intends to go where no one has gone before!

Its largely non-technical books take several different approaches. Journey with their authors as they

- Indulge in science speculation – describing intriguing, plausible yet unproven ideas;
- Exploit science fiction for educational purposes and as a means of promoting critical thinking;
- Explore the interplay of science and science fiction – throughout the history of the genre and looking ahead;
- Delve into related topics including, but not limited to: science as a creative process, the limits of science, interplay of literature and knowledge;

Readers can look forward to a broad range of topics, as intriguing as they are important. Here just a few by way of illustration:

- Time travel, superluminal travel, wormholes, teleportation
- Extraterrestrial intelligence and alien civilizations
- Artificial intelligence, planetary brains, the universe as a computer, virtual worlds
- Non-anthropocentric viewpoints
- Synthetic biology, genetic engineering, developing nanotechnologies
- Eco/infrastructure/meteorite-impact disaster scenarios
- Future scenarios, transhumanism, posthumanism, intelligence explosion
- Consciousness and mind manipulation

Herbert W. Franke

The Orchid Cage

A Science Fiction Novel

 Springer

Herbert W. Franke
Egling, Bayern, Germany

ISSN 2197-1188   ISSN 2197-1196   (electronic)
Science and Fiction
ISBN 978-3-031-60498-0   ISBN 978-3-031-60499-7   (eBook)
https://doi.org/10.1007/978-3-031-60499-7

English translation of the German edition "Der Orchideenkäfig" published by Goldmann, Wilhelm, 1961.

Foreword by Dietmar Dath

Something worse than what we think is (still) going on
On Herbert W. Franke's "The Orchid Cage"

This unique book paradoxically marries the urgency of prophetic terror to the calm sobriety of true wisdom. Central characters die casually. This sort of death is shown to be of little consequence; it can be survived. But that astonishing fact entails a cheapening of life as such. In complicated ways, it also acts as a crushing weight upon human curiosity. The central scenario plays out a slightly frivolous game. Two groups of humans compete with each other in exploring a seemingly empty city in a faraway world. The team that first reaches the center of this monument to a vanished intelligent species wins the competition. Though clearly up for anything, these adventurers seem at the same time strangely vacant. Their affect fall flat, their behavior makes the reader think of a person who is just going through the motions, adhering to ancient psycho-sexual scripts or algorithms like "who gets the girl?" or "what constitutes an alpha male?". The backdrop that these people have chosen for their high drama and low comedy challenges their internal settings in alarming ways. There are secrets here that stir self-doubt in those who think that the discovered world must be easy to unlock. Instead of conquering the relics of an apparently dead civilization and making them their own, they encounter its history in layers—some of it has a medieval appearance, but there are also houses like raindrops and spectral transportation systems which

v

match or even surpass anything the explorers know from terrestrial experience. The same holds for us as readers and inhabitants of the twenty-first century. In June 2023, Microsoft CEO Satya Nadella attributed to certain systems of artificial intelligence the power to "compress the next 250 years of chemistry and materials science progress into the next 25." The final challenge that Herbert W. Franke's protagonists encounter in his most provocative book is the legacy of such pronouncements. In the novel, machines turn out to be the caretakers of what remains when an intelligent species does not create its own progress anymore. This progress then becomes something that the society which has originally given rise to it can no longer keep up with.

Herbert W. Franke was by no means a Luddite. Equally well versed in physics or chemistry as in psychology or philosophy, he was one of the German-speaking world's most important science fiction writers in the twentieth century. In addition to that, he can justifiably be seen as one of Central Europe's pioneers of computer art and certain other kinds of interfaces between various STEM domains and the liberal arts.

The degradation of the human intellect, the threat to the physical integrity of sentient beings, and the distraction of the spirit that together form the uncanny subject matter of *The Orchid Cage* were well underway when the book was first published in 1961. But a warning that was not heard at the time does not become less relevant with age. Quite the opposite: as long as you are able to read a book, any book, there is still time. Franke's warning, meanwhile, has not just become more urgent but also more eloquent with age. It's as if history itself has activated parts of its voice that were inaudible before.

The warning rings as true as ever. Read it. Heed it.

Frankfurt, Germany   Dietmar Dath

Contents

The First Attempt

The first few seconds are always thrilling when consciousness awakens on a foreign planet. Right on cue, the image begins to build up, bit by bit, as if emerging from nothing. Detail lines up with detail, sometimes in waves, softly, sometimes abruptly. And above all, there is the tingling anticipation of something unimaginable: perhaps something superhumanly powerful, perhaps something cruel, even deadly.

The first thing Al sensed was the smell of thyme. He lay in a cradle of fragrant scent, embedded in a softly rustling and shapeless twilight. Gently, the feeling of gravity seeped through him. He felt as though something was lifting him up, and carrying him, then let him drift down again. A comfortable warmth spread over him. From the rustling, a soft rattling noise detached itself. The twilight turned orange-red. As if lifted by the wind, the last veils dissolved. The wishing and questioning awoke.

Al sat up. With a feeling of satisfaction, accompanied by just a touch of disappointment, he realized that he felt comfortable. There was no scorching heat, no electrical discharges, no sniffing dinosaurs. There was no obvious danger. His initial apprehension left him and he began to look around attentively.

He sat next to the barrack on a foam rubber mat. The rattling sound came from the door of the machine room. A robot trolley rolled over the leveled surface, a welding torch in its grippers. The shadow of the relay antenna, cast like a mesh over it, appeared to crawl in strange distortions up the automatic apparatus and down the other side.

Al was still fighting a certain lethargy. His muscles ached a little each time he moved and a dull pressure lingered inside his brain. But with each breath,

H. W. Franke, *The Orchid Cage*, Science and Fiction, https://doi.org/10.1007/978-3-031-60499-7_1

Al felt more alert, freer, and eager to act. He stood, squatted back down, then stretched up high. He unbuttoned the collar of his yellow bush shirt and took a deep breath. The place is good, he thought. Don has made a good choice.

The barrack was about a hundred meters above the valley. Right next to it, the machines had erected a transmission mast. Fifty meters to the side of it, at the same height, stood the hangar for the helicopter. But most striking of all was the crater from which the raw materials had come. It sat in the ground like an open wound, its edges bulging and blood-red where surface layers were exposed by weathering, its interior gray to black. Behind the barrack rose the mountain range, and below, a landscape of a thousand flat-topped hills and a thousand small lakes, stretching out under the orange-red glow of a foreign sun.

The sliding door of the barrack living room slid open and a cart drove out with Katja's limp body in its back seat. Using dozens of sensitive grippers, the robot cart laid the girl on the mat where Al had awakened only a few moments earlier. Now, Katja was about to regain consciousness. A slight twitch ran over her limbs, then another.

"Where is Don?" Al asked.

The robot cart stopped. "He's still sleeping," its speaker announced.

Al waved it away with a gesture of the hand. "Carry on!" he ordered.

So, Don was still sleeping and Katja was just waking up. Don and Katja had been approved as a couple by the Commission for Genetics and would one day have a child. This had brought the two of them together. Don had introduced Katja to his circle of friends. She had proven to be a good companion and had gladly taken part in all their games. The only thing that bothered Al was that Don treated her as a kind of possession and took the liberty of giving her orders. With a slight sense of schadenfreude, he imagined how Don would react at being the last to wake up.

"Hello, Don!"

The voice was so soft that Al barely heard it.

"Kat!" he called. "It's me, Al!" With a quick glance, he made sure that his khaki tropical jacket fit well and brushed back his dark blonde hair.

Katja was still on the mat and trying to sit up. Al knelt down next to her, slid his hand under her back and supported her. She blinked through her eyelids.

"This light is awful," she whispered.

"How are you?" Al asked.

"I've made it through. I think … I'm doing quite well."

"You'll get used to the light," Al explained. "Soon you won't notice that it's orange anymore. It'll go white. Everything will have normal colors. Then it won't look much different here than on Earth."

"But the sky," said Katja. "The sky."

"Even the sky will turn blue – just give it time. It'll happen all by itself. It's like a zero point shifting on a scale."

Strange, thought Al, the dirty gray color of the sky didn't bother me at all. Perhaps women perceive things differently.

He watched Katja. Her blonde hair was waving slightly in the breeze. The girl was still a bit pale, and this paleness emphasized her slightly protruding cheekbones, giving her an exotic appearance. Her dark blue eyes were half hidden under her eyelids. She wore red jeans and a black leather jacket over a black sweater. She was recovering quickly. Her movements became more purposeful, her gaze clearer.

"Great, Al," she said, "great that I can be part of this! It's my first time. Don't be angry if I'm a bit clumsy." She smiled, a little embarrassed, and seemed even nicer to Al than before. "But what about Don?"

She tried to get up again, but Al held her back.

"Rest a bit!" he said. "I'll check." He went over to the barrack and through the door. Compared to the brightness of the outside world, it seemed dark there. He felt for a switch and pressed it. A dazzling blue-white light flooded the room. Incredibly primitive here, thought Al, and this impression was confirmed when he began to look around. Everything was designed to save space. To the right, a window offered a view over the valley. Below it was a table and three stools. The door to the engine room was cut through the front wall, while the rest of the area was divided into compartments containing the most important tools. Everything not immediately necessary had to be specially made by the machines. Three beds were mounted one above the other along the left wall. Don was lying on the middle one. His heavy thickset body had sunk deep into the air mattress. The blanket reached up to the sleeper's mouth. Only the nasal region of Don's rather craggy face was exposed; his forehead was hidden by a short fringe of dark brown hair. He was already breathing.

Al opened the door to the engine room and called for one of the machines. "When will Don be ready?"

"In four minutes." The answer was prompt and precise.

"Then bring him outside!"

The machine obeyed. Al opened the door for it and followed. "He'll be ready soon," he said to Kat.

She had recovered by now. While the two of them waited next to the mat where Don was now struggling for consciousness, Katja began to ask questions. Everything seemed interesting to her now. All the new things that awaited her were still far off and somehow intangible; for the moment, she was on an island of human origin and human imprint, but around her lurked the foreign, the mysterious. As Al had predicted, the gray hue of the sky had disappeared – it was now glowing deep blue – but to Katja this blue seemed unfamiliar, just like the green of the plants and the brown of the rock face above them. She could no longer actually see it, but the sky was different, and the plants were different, and the rocks were different than on Earth. She could feel it, though. And the animals? She looked around, but there were no animals to be seen. Tomorrow, she thought, tomorrow! A warm and gentle breeze was flowing up the slope from the lowlands and bringing with it waves of a scent that smelled like thyme. But it was most likely something quite different, the very breath of mystery.

"Hey!" Don called out, but his voice was still weak. He struggled to prop himself up on his elbows. "Are you guys okay? I'm still spinning. It's shameful, isn't it? How are you guys doing?"

"Pretty good," said Katja. "How do you feel?"

"Getting better. It's not important. Have you seen anything of the others?"

Katja stroked his forehead.

"Nothing, Don. You haven't missed anything. We've only been awake for a few minutes."

"That's good then." Don sighed and fell back onto the mat. He crossed his hands behind his head and closed his eyes. "Tell me. What's it like here?"

"Pretty benign," said Al. "Green hills, mountains, lakes. The air is good, the temperature pleasant. Nothing particular to report. Let's hope it won't get too boring!"

"Hardly likely," said Don. His eyes were still closed. "The others will take care of that. And then there's the old city!"

"Where is it actually?" asked Katja.

Don sat up. He wore ankle-length corduroy pants and a tight velvet jacket with gold buttons. He already looked significantly fresher. "Today we'll check the supplies. And tomorrow we start." He peered down at the hilly landscape below, and across it to the horizon. It was evening. The red disc of the sun hung in the haze. And as it slowly sank lower, a distant line began to glitter. Red sparks glowed in the distance, then disappeared.

Don raised his hand and pointed: "There is the city."

* * *

By the next morning, they had all recovered from the after-effects of the transmission process. They stepped out of the barrack building and stood in the dawning day. Once again, they were enveloped in that intoxicating scent of thyme. They looked over the hilly countryside before them and felt drawn by some powerful force that emanated from those untouched expanses.

"What are you waiting for?" Don called. "Come on, let's fly. There's no time to lose!"

"Are we going to visit the city right away?" Kat asked.

"That would be the nicest thing," said Al, "but shouldn't we rather follow the tried and tested rules? The investigations. The composition of the air. The chemistry of the soil. The spectrum of electrical waves. The zoology and the botany …"

"Oh come on! The air is breathable – we don't need a chemical analysis for that," cried Don. "And there's no sign of any animals. Do you want to stay here and pick flowers?"

"Don't be boring, Al," Kat pleaded. "You still have plenty of time for the plants. Come with us – to the city!"

"You know what this is all about, Al!" Don shouted. "We have to beat the others! We have to be the first!"

They walked over to the helicopter and climbed up. Don sat in the pilot's seat and started the engine. Katja and Al took their seats behind him. The whirring of the propeller blades turned into a deafening roar, and the aircraft lifted off the ground. They ascended to a height of five hundred meters.

The helicopter's transparent canopy allowed a clear view in all directions. The landscape beneath them was like a scene in a children's picture book, toy-like and lovely. One could have taken it for a natural park in Finland, if it had not been overshadowed by the huge mountain range that arched around them. From their elevated viewpoint, they could clearly see the peaks and ridges of the mountains, which now appeared in the sunlight as a dark, jagged strip above the lowlands.

"What are the chances of discovering something here?" Kat asked.

"Good question," Al said, "you haven't told us anything yet. Stop being so mysterious!"

Don turned the nose of the helicopter towards the target and they began to move in the direction of the city. He pushed the throttle hard, and the hills began to slip away briskly beneath them.

"So, listen up," said Don. "Jak and I, we found the planet together. We were planning to look around the Magellanic Clouds with the synchronous beam mirror. I don't know how it happened – maybe Jak had set the distance wrong – but we were looking into an empty region and were about to switch, when we stumbled upon a small isolated star. Attached to it was a giant Neptune-like planet. Then we looked further inside the system and, well, this is the result!" Don pointed down at the ranges of hills shifting against each other like conveyor belts.

"Did you find the city right away?" Kat asked.

"There are several. They were not hard to find. The planet is ninety percent covered with mountains. The remaining ten percent of green spaces stands out clearly from the brown and gray. All the cities are located below, in the hilly lowlands. But that one," he pointed forward with his chin, "that one is the largest."

A hill was now growing over the horizon in front of them. It was a hill that differed significantly from the others. Against the blue of the sky, its many tiers and outcrops of towers, punctuated by notches, cut a quite different skyline from the gently curved and smooth silhouettes of the other mountains and peaks.

"Could there be anything intelligent living there?" Katja asked.

"No way," Don replied. "The time required for organic evolution is much too long. The planet is not old enough for intelligent life to have evolved here twice. There's hardly any chance of that happening. So far, no one has encountered intelligent beings here, although traces are often found. No, the city is certainly dead. It has partly fallen into disrepair."

Al had raised the binoculars and pointed them at the hill covered with buildings. He had to agree with Don. What from a distance had looked so imposing, like something from a fairy tale, and which had appeared yesterday in the grazing light of the setting sun like some pristine golden city, awakening impressions of legends and dreams, turned out to be a barren, crumbling castle complex.

Kat turned to Don again: "But what happens to intelligent creatures when they have reached their highest level of development?"

"They kill each other," Don explained. "This is what makes our own culture different from all others. And although it did actually start with us, with bacteria and nuclear energy, we stopped the war in time."

By now, some features were already visible to the naked eye: towers dotted with patterns of window openings, curved bridges over street canyons, scaffolding constructions, and masts, but much of it collapsed, twisted, falling into disrepair.

Al lowered the binoculars.

"Why didn't you come here with Jak?"

Don laughed.

"Do you think I care about that kind of old junk? The study of technical achievements? Space archaeology? What a lot of nonsense. I'm here to experience something. Something entertaining and exciting. Of course, Jak too."

"And that's why you ..."

"That's why each of us formed a group. Yes. Whoever first finds out what these beings actually looked like wins the prize. Each group has to work independently, and neither may interfere with the other. But knowing Jak ... If he gets the idea that we might snatch success from under his very nose, he'll do anything he can to prevent it. That's why we have to be careful. So, keep your wits about you. What we're trying to do here won't be easy!"

The landscape beneath them was still a patchwork of blue and green. The lakes were scattered across the grassland like so many fallen leaves. From time to time, strange rock formations stuck out. The area around was dotted with bushes.

"What are those?" Katja cried out suddenly. "Al, pass me the binoculars!"

He handed them to her. So far, he had been focusing on the city, but now he turned his attention to the area beneath them. The slightly undulating surface was covered with fresh green meadows. But the green was not completely uniform. Al noticed areas where it appeared a little paler and revealed certain regularities. There were stripes of lighter green, mainly straight, but occasionally curved, branching off in different directions. Al recalled an archaeologists' trick using the shadows cast across a landscape to detect the relief due to artificial changes, but the sun was already too high. Nothing definite could be deduced.

The features that Katja had noticed were like dabs of color, some pitch black, some tinged with green.

"They're holes in the ground," the girl shouted.

Al took back the binoculars and confirmed: "It looks like impacts from projectiles, like shell craters. Could they be traces of a war?"

"Probably," said Don. "They must have killed each other somehow."

But Al shook his head. Speculations were not enough for him. He would have preferred to deal more carefully with such questions, but he dismissed his concerns. After all, it doesn't matter, he thought.

They were now flying in the immediate vicinity of the city. The lighter green stripes were beginning to condense into a network, strengthening Al's suspicion that they were paths and roads. But why are they so overgrown with plants, Al wondered, while the impact craters are still bare?

From the pilot's seat came a curse. Surprised, Kat and Al looked at Don.

"Something seems to be wrong with the steering!" Don complained. "The machine keeps veering to the side."

"Just as we're coming over the city!" Kat looked straight down through the glass bottom of the cockpit – the first buildings were arranged in a regular pattern of white dots across the green areas.

Al was watching Don. It was indeed strange. The machine had difficulty maintaining its direction, like a car on a steeply banked road. And if Don overcorrected, the helicopter swung to the other side.

"It's not the controls," said Al.

Don snorted irritably.

"To hell, what then?"

"Try flying more tangentially, as though you were going around the city in a circle. Nothing is stopping you when you do that – do you notice it?"

"Why, yes, it looks like you're right!"

Katja was following these attempts without understanding. "What's going on?"

"Something is diverting us from our course toward the city," said Al. "Go into automatic, Don!"

Don pressed the red button and it changed to green. Then he moved the course indicator around slightly. They waited anxiously … the direction stabilized.

"So that's what we have to do," said Don.

"Well, I thought it was," said Al.

Don gave a worried glance toward his companion.

"Look down!" said Al.

"Well?"

"We're not moving."

"Damn it all, you're right!"

It was not immediately apparent from above, but looking down for a while, it became clear that their position was not changing. The landscape stood motionless, the helicopter hanging above it.

Don pushed the throttle down.

"Maybe there's a counter current?"

Indeed, the machine slid forward a little.

Don revved up the engine, the propeller roared, and again the craft pushed forward a few meters. Then it stood still for a few seconds and suddenly began to sway wildly, like a wood drill that has hit metal.

It swayed so much that the three of them struggled to stay upright.

Don reduced the pressure on the throttle and the rocking motion stopped. The machine began to drift backwards, tail first, a few dozen meters away from the center of the city.

Don cursed angrily under his breath. He flicked the automatic control button off and the helicopter promptly swung round a hundred and eighty degrees. Al and Kat were pressed against the side wall, while Don clung to the control stick. They accelerated, flew back a little, then curved back around and flew at full speed toward the city.

"No, stop!" Al shouted, but the machine was already rearing up, then sliding sideways, spinning slightly. Then again, Don went into full throttle and curved back through a half circle.

"I'll get through, even if the whole thing falls apart!"

Again, the centrifugal force tugged wildly at them.

"Please don't, Don!" Katja cried out.

But Don wasn't listening. He increased the power and it was as though the helicopter was being thrown at full force onto a springy cushion. There was an ugly crunching sound in the framework and they began to spin downwards, the fall turning into a kind of sideways sliding motion. Don regained control of the flight.

Kat laid her head on Al's shoulder, shaking gently. Don didn't say a word. He steered back to camp.

* * *

Not even twenty-four hours had gone by and they had already suffered a defeat. Don threw himself angrily onto his bed, and Katja sat listlessly beside him.

Al carried the inspection case down the slope. When he reached the plain, he set it on the ground and began his routine physical and chemical examinations of soil and rock samples, then turned to the plants. He collected samples of flowering plants, grasses, and mosses, setting aside for preservation any species that differed significantly from those on Earth. However, the results were meager.

After a while, Don and Kat joined him. Don was still in a bad mood. He stomped around aimlessly, occasionally plucking a flower and pulling off its petals.

"Al!" he said after a while. Al, who was bent over a microscope, muttered something unintelligible in response.

"Have you found anything interesting?"

"Nothing special so far."

"Do you know what strikes me?"

Al was tapping on a boulder with a geologist's hammer and collecting the fragments.

"No. What strikes you, Don?"

"Everything's a bit too dead here for my liking. There are no birds, no quadrupeds. No ants, no flies. Not even a flea. Have you seen any animal life of any kind?"

"Take a look down the microscope, Don. What do you think?"

Don looked into the eyepiece.

"What's that?"

"A drop of water. I stirred up a bit of soil with water and that's a drop of what I got."

"So? I don't see anything."

"That's just it." Al pulled the slide out from under the clamps. "Nothing. Not even microorganisms. There must have been some kind of evolution here, from the lowest forms of life to the highest. If there wasn't, where did the builders of the city come from?"

Don had no answer.

"I want to show you something else. Maybe it'll change your opinion that routine analyses are a waste of time. Do you know this formula?" He handed Don a piece of paper. "It's the result of a chemical analysis. Do you know what it is?"

Even as Don shook his head, Al continued: "It has no proper name, but it's extremely effective at what it does. Watch this. I'm going to show you a little experiment."

He slipped a polypeptide slide into the microscope's vacuum chamber, set it to ion-optical magnification, and asked Don to look at the image.

"Okay, I see long stretched-out rectangles. What's that about?"

"Bacteria from our biological test material."

Al dipped a glass rod into a test tube and brought out a crystal-clear drop of liquid. He pulled out the slide, wiped the drop off on it, and pushed it back into the beam path. He then looked through the eyepiece to make sure that what had happened was indeed what he had expected. Then he signaled to Don to take a look himself.

"What do you see?"

"The rectangles are dissolving into shreds. So, what's going on, Al! What does this mean?"

Al bent down and tore a few leaves from the grass mat they were standing on.

"According to the formula, it's a kind of antibiotic. It's unusually strong and it works very fast."

"Where does it come from?"

"It's everywhere. There's a thin coating on the plants, on the rocks. It's dissolved in the water. It flies through the air like dust."

Don shrugged, bewildered.

"So, how do you explain that?"

"I can't explain anything at all. For the moment, I'm just making an observation. And now there's another surprise." He showed Don some fragments he had chipped off a larger stone.

"It's a sample from over there," said Al, pointing to one of the steep cliffs out in the hill country. "I've already had a quick look, but just to be sure, I'm going to do a full spectral analysis."

Don and Katja watched the deft movements of his hands. Using a piece of rough platinum foil that looked rather like a nail file, he scraped some powder off the sample. Some of it went into a rotating crystal X-ray diffraction device, and he put the foil itself into a spectrograph. Then they all stood waiting for the results. Don was clearly impatient, while Katja looked on with uncomprehending astonishment, and Al tried to appear calm. After a few moments, the transport wheels clicked, and the tape came out, two ten-centimeter-long streamers, snaking their way into the daylight. Al reached for them and smiled.

"Don't leave us standing around like idiots," Don shouted. "What have you found?"

Al was still smiling.

"Plastic. The rocks are made of plastic."

They were silent for a while. Then Al said: "There's no point in burying our heads in the sand. There are plenty of things here that are puzzling. Germ-free soil and artificial rocks – okay, so it doesn't fit any reasonable picture. But there's nothing bad about it. The most worrying thing is perhaps the incident over the city. But even there, there's no particular reason for concern. After all, you didn't have to smash up against the resistance like that!"

"Oh, Al, stop it!" Katja pleaded. Don scowled.

"Why shouldn't I talk about it?" Even the usually imperturbable Al was now showing signs of annoyance. "The task we set ourselves isn't going to be as easy as we thought. If we want to go on, we have to stop acting like children. Unless of course you want to give up?"

Katja looked at Don, who acquiesced.

"Me neither," said Kat.

"Alright then. Let's proceed a bit more systematically. We've established some facts that we don't yet understand. In the future, let's take every opportunity to gain more information. I'm convinced that everything can be explained rationally!"

Don had sat down and was twisting a few blades of grass in his fingers.

"Do you think this antibiotic could harm us?" he asked suspiciously, peering at a little dust on his fingers.

"Calm down," said Al. "It can't harm us, you know that yourself."

Don touched Kat's neck with a blade of grass and laughed when she started back suddenly.

"You're right, Al," he said. He had regained his usual good humor. "You're right. So, what are we going to do now?"

"Best if we finish the tests. In the meantime, we can think about how we might get into the city."

"Do you think we can get past the barrier?"

"What was it anyway?" Al asked. "A current of some kind? A strong wind? A force?"

"Yes, something like that. Anyway, it wasn't hard. Maybe a soft invisible object?"

"I don't believe in invisible objects. I would say it's rather a force – a kind of antigravity, perhaps. But that's not the most important thing. More important for us would be to know what purpose it serves."

"Well, that's clear, isn't it? The city has been cordoned off. No one is allowed in. Perhaps something still in place from the distant past."

"Maybe. But another possibility occurs to me. I can't say I really believe it, but couldn't Jak have … ?"

"Jak? Hmm – that would be typical of him. But how could he have done it?"

Suddenly, something strange began to happen. Barely noticeable at first, a bright high-pitched ringing sound began to shake the air. Then, a white-hot ball shot down from above like lightning making a terrible screeching noise and ended with a dull thud. A moment later, everything was as before, except for one thing: on the side of the next hill just opposite them, a rather inconspicuous bump in the ground, a new black crater yawned.

The three onlookers were pale. Their hearts were pounding. They stood hunched over, as if someone had just set upon them. It took them a while to recover. Then Al cried out, breaking the silence: "I've got it!"

"What?" asked Don hoarsely.

"Tomorrow we'll enter the city!"

"What? How are we going to do that?"

"What we just saw was a meteorite. The craters are meteorite craters. Now, think back. Did you see any near the city?"

"I can't remember."

"Well, I do! There were none. And now I know what the repulsive force field was for. It's a defense against meteorites, which are apparently commonplace here."

"That makes sense. But what does this have to do with visiting the city?"

"Isn't it clear?", asked Al. "They only need to cover the city from above to protect it. I'm sure there'll be a way in under the shield. There must have been paths leading out of the city in the past. We have to approach it at ground level, then we'll be able to get in!"

Don clapped Al on the shoulder.

"Right, this funny force field is not really directed against us! So, nothing stands in our way now!"

Al nodded confidently.

"Looks like it," he said.

* * *

The next day they set off again in the helicopter. Don approached the city at low altitude and landed when he noticed the first signs of resistance.

"You were right," Don said to Al, "further inside there are no more craters. Hopefully, you were also right about getting in unchallenged at ground level!"

They got out and began to walk slowly towards the city. Their steps were uncertain. Involuntarily, they stretched out their hands as though they were blind. Somewhere above them, perhaps also in front of them, was something invisible yet effective, the first obstacle on their journey into the unknown, the first sign of an unknown power and proof that they were in the presence of a high level of technological superiority. It hung in the air above them, hidden from their senses and yet real. No one knew what transformations and reactions it was capable of. So far nothing had happened that had been specifically directed against them. Meteorite falls were an apparently well-known and not particularly unusual natural event here. But now, as they approached their goal, they realized for the first time that they were entering another realm, an inscrutable world, the sphere of a spirit that was long gone, but whose legacy had not yet perished. What they were doing now was different. It was much more than they had done before, and it was significant in a way they could not have anticipated.

Feeling small and hesitant, without the help of any technology, relying on themselves alone, they advanced cautiously over the grassy terrain, which had never been a human territory. The helicopter was now far behind, suddenly much further away from them than they had thought. But Al had been right! Down here there was no barrier. With every step, they gained confidence, their assurance increased, and at last they lowered their probing hands. As they looked at each other, there was a trace of pride and shame at the same time.

From down here, everything looked quite different from the bird's eye view they had had before. Any resemblance to earthly neighborhoods had greatly diminished. Only the loose arrangement of the buildings was still reminiscent of a residential suburb.

They were soon standing in front of the first buildings and could get a better view of their shape and the way they were arranged. The vast majority of them had smooth, rounded frontages made from some kind of reflective, iridescent material. Viewed from the city center, these facades pointed radially outward. The front of each building was its highest part. They lowered and narrowed to the same extent towards the back, so that the ends tapered to a point toward the city center. Over their highest points were shimmering metallic arches stretched across with nets.

"They look well built," said Don. "There isn't a single window. They look a bit like bunkers. What do you think?"

"Maybe they're storage rooms," Al suggested. "It's hard to say."

Katja looked on with wide eyes.

"Chic," she said. "Like modern garages."

"Everything is still well preserved here," said Al. "The periphery seems to be the newest part of the city. What you can see there in the interior doesn't look so new anymore."

They walked between two of the elongated buildings. The grass was lush, as in the open country, but there were narrow grass-free strips along each house. Katja took a few steps to one side to reach one of them.

"The roads are bad here," she said.

"I noticed that too," Don confirmed. "The people here probably didn't use roads at all. I don't know how …"

A whistling noise and a scream from Kat made him fall silent. He spun around and saw Al taking a few steps forward before stopping. But Kat was gone.

"Where's Kat?" he shouted.

Al was struggling for words, but all he could do was point to the wall of the building. Don ran towards him and jabbed him roughly in the side.

"What happened, Al, answer me!"

"She was there just now, and suddenly she screamed. There was a kind of dark opening in the wall, and I just saw it closing up. That's all."

"Where was the opening?"

They ran over to the building and examined the wall. Al knew roughly where the hole had been, but when he felt the wall, he found no obvious trace of it.

Then the whistling noise he had heard before sounded again, much closer by. The wall in front of him seemed to open up and he felt he was being pulled inward. He tried in vain to hold on to the edge of the wall, but he was losing his balance, and something gently pushed him into a sitting position. There was a warm yellow light. He began to move up a gently inclined and softly curved spiral surface, with the yellow glow running ahead of him. Images appeared before him like those in a kaleidoscope: patterns of color, cacti, keyboards, wire sculptures. Then the elevator stopped. In front of him lay the wide plain with its hills, lakes, and rocky outcrops, and beyond, the speckled brownish black wall of the mountains with their glittering crown of icy peaks and glaciers. Next to him sat Kat, her face hidden in her hands.

Al forgot the magnificent view. All he could think of was Kat. He tried to get up, but something was holding him back, gently but unyieldingly. He leaned over to her as best he could, placed his hands on hers and tried to soothe her. The trembling stopped and she rested her head on his shoulder.

Suddenly, Don's voice boomed out: "This is going too far, Al! Have you gone mad? Kat, what are you doing? Don't forget what we agreed – or I'm out!"

"It's okay, Don. She was just really scared."

"There's no reason for that. She knows very well what we're doing. She doesn't need your comfort!"

"Sure, Don. I didn't mean any harm."

Al was leaning back again. They sat side by side in springy lounge chairs, red in the face, staring blankly ahead. Before them, almost within reach, lay the paradise of hills and lakes, and the majesty of the mountains. The shapes seemed to have incredible depth, as though nothing separated them from their onlookers. The sun lit up glittering chains on the ridges, while the landscape below was resplendent in a harmony of color and light. The curved blue sky seemed almost real enough to touch, and in this sky the stars were visible. But it was not just what they could see: wind-stirred grasses rustled softly and a blanket of warmth lay comfortably over their limbs. A faint scent of thyme wafted over them and conjured up a hint of vastness, timelessness, and freedom in the room.

The three of them stared blankly ahead, seeing nothing, hearing nothing, feeling nothing. They were separated from these things by more than a wall of strange glass, by more than millions of light-years, by more than millennia of progress.

Al was the first to recover. He had by now regained some of his usual composure, but not the calm of those who might have sat here an unimaginably long time ago. He looked up and said, "The land at the edge of the mountains. The iridescent wall is a window."

"It's more than a window," said Katja. "You don't just see the landscape – you can hear and smell it too."

Don tried to stand up, but was no more successful than Al had been earlier.

"It's not a window. It's more like a movie screen, or a monitor. Anyway, can someone tell me how to get out of here!"

In his attempt to stand up, he stretched out a hand to one side. Without warning a control panel grew out of the ground next to him. He pressed a button. Katja screamed. The backs of their seats began slowly tilting back, pulling the upper parts of their bodies with them. They lay there looking up at the ceiling, unable to get up. Don fumbled again for the switches, and moist, warm mist droplets began to spray down on them. Their hands and faces tingled pleasantly under their effects and a sharp and invigorating smell drove away the scent of thyme. Don fiddled with the control knobs again and the light went out.

"Stop," Al shouted, but too late. Something pliable yet strong was touching him. It gripped his muscles, kneading them, flexing them, and rubbing. Something circled around his forehead with just bearable pressure, but still he couldn't move. He tried to flail and kick, but the invisible hands wouldn't let him go. Still, they pressed, stretched, massaged.

Finally, the ordeal was over. A violet twilight glowed. He was tired, unspeakably tired. He gave in to the tiredness and the swaying masses beneath him.

"Al … !"

Again: "Al!"

With difficulty, he tore himself away from inexpressible dreams.

"Hey, wake up! We have to get out of here!"

Al turned his head: there lay Don. He turned the other way: there lay Kat.

"Yes. We probably should try to get out."

What a pity, he thought, I would prefer to stay here. Just do nothing. Not try to do anything. Just dream.

"But we have to get out of here."

On the other hand, maybe they won't let us go. After all, why go outside when you are safe inside? But we have a mission to accomplish.

"Don, there must be a button for that. But please don't set the massage again."

Surely, there were also buttons for food, for music, for long-distance calls.

"Isn't there a single button all alone at the bottom of the panel?"

The most beautiful thing was undoubtedly the deep sleep filled with vague dreams, the overwhelming sense of fatigue. Al's eyes were still half-closed, but he noticed that Don was now reaching for the control panel again.

Then the yellow light was back, running just ahead of him. He felt himself sliding gently downwards. A bright rectangle opened up before him and he was greeted by a cold and dazzling light.

Al was standing in front of the building again, with Katja leaning against the wall next to him. There came the now familiar whistling sound from behind and Don staggered out.

* * *

The houses brooded silently in the sunlight, white teardrop-shaped structures on green mats. The branches of the bushes swayed gently in the breeze, petals detaching from them from time to time and floating down feather-light under the deep blue sky.

"What now?" Al asked.

"What now, what now!" Don repeated. "We keep going. What else?"

"And where are we going?"

"To the center, of course. This is not a holiday!"

Al said calmly: "We entered the building unexpectedly and came out again unexpectedly. I assume that the inhabitants of this city would have stayed in these buildings, in the large room behind the front wall, which looks perfectly opaque from the outside. They would have sat in the lounge chairs, slept, ate, and drank, had massages, and in-between they would have watched the picture on the screen. I'm sure they liked the view of the hill country and the mountains, but they would probably have wanted to see something else from time to time. Presumably, they could also set up plays, shows of different kinds, and films using the control panel, and in plays and films, people would appear. Or whatever they may have been. We should start by taking a closer look at that apparatus. We could already be close to answering all our questions."

But Don was already moving towards the city center. Half over his shoulder, he called back: "Do you want to be sprayed again? Do you want another massage? Do you want to be sent to sleep again?" He brushed a handful of flowers from the branch of a low bush impatiently. "Do you want Jak to get ahead of us?"

Katja stood there undecided, watching Al to gauge his reaction. He glanced back at her.

"Oh, come on then!" he said resignedly.

They wended their way between the buildings, wisely keeping themselves well away from their treacherous walls. They all had the same shape, like a teardrop tapering toward the back, and they were all made from the same white or ivory-colored material. Still more of their bulbous, mother-of-pearl front walls came to meet them as they moved forward, reflecting the blurred sun behind them. Only occasionally would they come across a gray, block-shaped building, something Al would dearly have liked to examine more closely.

He noticed that Katja was getting tired and suggested to Don that they take a break.

"Well, if you insist," grumbled Don, who was actually ready for a rest himself. In the meantime, Al had noticed a peculiar grey building nearby, different in shape from the others, and he approached it cautiously. He couldn't see any openings, but he already knew that this didn't necessarily mean there was no door. He walked slowly around the building. Then suddenly, a gate opened next to him, and he was grabbed by what seemed to be an invisible force and pushed onto a seat that whisked him inside. This time, he was not so surprised.

The same yellow light accompanied him again, but this time the path was short, only a few meters long. He arrived directly in a chamber that looked like a miniaturized version of the room he had been in before, and came to a halt in front of a circular screen about the same height as a human being. The surface promptly lit up and an image appeared: it showed the familiar white buildings in the neighborhood. Al immediately realized that it was the view from the cuboid building he was actually in at the moment. He looked to the side and found what he was looking for: a panel covered with levers and buttons. He hesitated for a few seconds, then pressed the button at the top left. There was a barely perceptible jerk and the image in front of him began to move. His sense of motion combined with what he saw to produce a reasonable overall impression: he was in fact moving. He didn't know how, nor where to, but he knew he was moving. Katja and Don appeared briefly in front of him, their frightened faces looming suddenly large, and he heard them shout out. They ducked and he glided over them, moving quickly further away between the buildings. With the utmost care, he began to operate the buttons and levers, carefully observing the effects this had. In this way, he soon learned how to zoom in and out of the image, accelerate, slow down, and finally stop the motion – or rather, the flight. At the very bottom of the panel was a single button that served to eject the vehicle's occupants in the now familiar way. He looked at the vehicle from the outside: it was like a gray cylinder, rounded at the front, without any outstanding details.

Al looked around to get his bearings, but there was nothing of particular interest. So, he went back toward the door and got himself carried back to the driver's seat, where he could examine the control panel in more detail. Next to the buttons and levers, he noticed a kind of sketch with a ring-shaped network of lines running over it. Three points were marked on it. Two of them, a blue one and a green one, seemed to be fixed in the sketch itself. Another was formed by a red plate that could be moved over the lines.

This gave him an idea. He set the vehicle back in motion. So that was it – the blue dot began to move. Now he pushed the red plate over the blue spot to the green one and waited somewhat anxiously. He was beginning to think he had been mistaken, but then the flying car turned a corner, and the blue dot did the same on the plan. Because it was indeed a plan of the city showing a road network. Al first had to find out how the vehicle was connected to the road system. Because it was clearly bound to it in some way. It was clear that the red plate could be used to set the destination. He thus moved it back to the starting point, and the vehicle duly went back there. He was delighted to find himself gaining control over this strange technological organism and surprised at how proficiently it obeyed the slightest movement of his hand. He began to go faster and faster, then braked abruptly, and yet without jerking and skidding, swept elegantly around corners, swung between blocks of houses, and darted along the longer straights at lightning speed.

Not ten minutes had passed since he had left his companions, when suddenly they came in sight again. He stopped just beside them and gave the signal that he wished to get out. The seat began to move, and as if carried by a wave, he was brought over to the door.

"Get in," he called, "you don't have to walk anymore!"

"That was quite a shock when you came rushing up and then disappeared. Where did you go?"

Al was in a great mood.

"On a little test drive!" he exclaimed. "Just get in. It couldn't be more comfortable!"

Don was the first to approach the door and disappeared inside. Katja followed, and Al was the last to enter. He explained the keyboard to Don, who set the vehicle in motion towards the inner region of the city.

* * *

Don also quickly learned how to operate the controls.

"I'm glad nothing happened to you," said Al, "when I passed over you earlier. What does it actually look like from the outside when this thing moves?"

Don was still focused on the road, even though the vehicle seemed to steer itself on its own. When he answered, he was facing towards the viewing panel.

"It doesn't remain on the ground. It flies, or rather, it hovers. About three meters above the ground. Like a zeppelin, but much faster. But don't ask me how it does it."

"Clever of you to find this, Al," said Kat. She was sitting in front of him and behind Don. As far as her seat allowed, she slid either to the left or the right, depending on where there was something to see. "Walking is boring, but this is fun!"

"I think there's some kind of guide rail buried in the ground; maybe this thing is connected to it by radar. It probably also gets its energy from there. In any case, it has to stick to the pre-drawn road network."

Hearing his remark, Don tried all the levers and buttons again, only to confirm what Al had said. After Don had accelerated, braked, and accelerated again several times, he gave up in his attempts.

"Alright, let the crate run. We don't have much further to go to reach the center, and anyway, that's probably as far as it'll go."

"This kind of roadless locomotion is actually ideal," Al noted. "The way we developed things on Earth has taken a quite different direction. The whole planet is cut up by road systems. I always felt it was a shame that no piece of landscape remained untouched."

No one commented.

After a few hundred meters, the vehicle stopped, and they were carried out in the usual way.

"Too bad," Kat remarked, "does this mean we have to walk now?"

"You can stay here if you want," Don replied. He peered around to get a better idea of where they might go next. Reality seemed pale and sober compared to the high spatial contrast and bright colors of the image on the screen. This impression was further enhanced by the fact that they seemed to have reached the inner edge of the modern part of the town.

"Now I know why the rail network only covers the outer ring and stops here."

He drew a ring in the air. "We've reached the parts of the city that were built before the zeppelin was invented."

Al nodded.

"The city seems to have developed radially outwards. They probably built that way, one ring after another."

"But why didn't they modernize the inner parts?"

"Why should they?" asked Kat. "They built their modern houses so that they always looked outwards. They didn't care what it looked like inside."

"Exactly," said Al. "And they were certainly long past the time when overpopulation was a problem."

"Why are we actually going further into the city?" asked Katja. "According to what you're saying, no one would have been in the center recently anyway. So, what are we going to do there?"

Don shifted uncertainly from one foot to the other. The interior of the city appealed greatly to his sense of adventure. On the other hand, he was also keen to reach their main goal first – and he would have been willing to do a lot to achieve this, even a meticulous scientific investigation. But then he would have had to agree with Al, who had already advocated a thorough examination of the buildings, and he couldn't easily decide to do that.

"We haven't seen any of the inhabitants of these houses," he said thoughtfully.

"Where did they all go?" asked Kat. The question clearly bothered her. "Your theory was obviously wrong!"

"What theory?" Don asked irritably.

"Well, that the inhabitants all killed each other. The houses of the last generation aren't even damaged!"

"What does that tell us? Maybe poison gas or bacteria?"

"We would have found traces of them."

"Maybe they took shelter down in their basements?"

"That's quite possible. We didn't check," Al pointed out.

Don hissed irritably.

"We didn't have much opportunity to do that. Or did you see a way to escape from your chair?"

"Do you really think we will find the inhabitants of this planet mummified in their basements?" Katja was torn between horror and curiosity.

"I think the problem is actually completely different," said Al. "It would be worth thinking more carefully about this."

"Are you trying to sell us more theories?" asked Don. He adopted a slight mocking tone, to assert his superiority, but secretly he admitted to himself that it was just an attempt to cover up his own deficiency.

"The question is," Al continued, ignoring Don's irony, "what do intelligent beings do once they have survived the phase of self-destruction? Surely, you don't really believe, Don, that we're the only ones to have managed to get past that stage?"

"What should they do?" Don asked in a condescending manner. He declaimed haughtily: "And if they didn't die, that means they're still alive today."

"Do you really think that a race that has achieved a certain technological level has nothing more to fear?"

"Maybe they all died out of their own accord?" Don now clearly showed the annoyance this discussion was causing him.

But Al didn't let up.

"So, you mean that, once they've reached the point where they've eliminated all threats, they can satisfy all their desires, and they have no further problems, then it becomes pointless for them to go on living. They'll want to just lie down and die. Isn't that a bit simplistic?"

"Leave it to me what I mean," Don shouted, and now he was really angry. "I suggest we try to get up this tower." He pointed to a high-rise block with a dome made of the same iridescent material that they had already noticed on the front of the earlier buildings. It towered over most of the other buildings. "From there we'll have a good view over most of the city. Maybe we'll find something useful, and then," he didn't hide how hard it was for him to make this concession, "then we can still examine one of the houses if it's really necessary."

"A good idea," said Al, and winked at Kat, because he thought that she must also be amused by Don's excitability, but she just looked at him with wide eyes and quietly followed on behind Don.

They were now leaving the world of order and cleanliness behind them. The buildings on the right and left moved closer together, separated by gaps that could be called streets. These houses were of various colors and probably made from different kinds of material. Some were large, some small, each shaped according to its builder's whim. And they were falling apart. Paint was peeling off the walls, spongy weathering products filled nooks and crannies, and shreds of a transparent fabric curled in long openings that must once have been windows.

However, there were also signs of destruction that were not due to age, but rather to something from the outside. Things like wide cracks in the walls, collapsed roofs, and charred ruins. In some places, there were even traces of repair work. Some of the cracks were filled with a kind of mortar, and makeshift roofs had been erected over some ruins.

Then they came across a funnel-like opening in the ground.

The ground here was devoid of undergrowth: no grass, no shrubs. And the path, covered with a mass of dust and rubble, suddenly came to an end. The ground broke off abruptly; the deepest spot they could see was about five meters down – a hollowed-out depression covered with yellow dust. There was yet more dust stuck on the funnel wall. Don knelt at the edge, bent down, and fanned a little of the light, yellow material aside with his handkerchief. Lumps of red, brown, and black slag came to light.

"A meteorite crater!" exclaimed Kat. "But I thought the city was protected from meteorites!"

"It is now," explained Al, "but it wasn't before."

"Hmm," grunted Don as he shook out his handkerchief. "I guess they invented the shield when they started to build the outer belt."

Al agreed.

"It was probably at just this point that they finally learned to control matter the way they needed to. I don't know much about technology, but I believe some things here are completely unknown to us. The shield, for instance. Or the mechanism that got us in and out through the doors when we entered the houses."

Katja looked uneasily up at the sky. There was absolutely no visible sign of the protective shield – and what one cannot see, one can easily begin to doubt. Don was still pondering.

"Could there not have been a bombardment by some hostile power? Maybe the system is automatic and it's still operating today because no one has turned it off?"

Al shook his head.

"I can't believe that. Such a power would have had a more effective means to attack, not just these relatively harmless projectiles."

"Well then, let's move on," commanded Don. "The way to the lookout tower must be here somewhere!"

From among the jumble of houses, it was not so obvious which way to go. They had to look for places from which they could occasionally catch a glimpse of the structure towering above the flat roofs to work out its approximate direction.

They turned a few more corners and suddenly they were standing at the foot of the tower. It too bore traces of decay, but it still seemed to be reasonably secure. In contrast to the modern houses they had encountered before, the door openings were easily recognizable.

"Be careful," Kat advised as Don stepped over the threshold. "If the transfer machine is defective, you could get stuck or crushed."

Don waved it off casually.

"Don't worry, kid!"

Nothing special seemed to happen. He plunged into the darkness, groped for a switch, and found none. Gradually, he got used to the poor light and began to pick out the outlines of the furnishings in the room. A sloping path led up to the left, while to the right there hung a box-like structure on vertical rails. Don suspected it might be an elevator and turned to examine a plastic panel with push buttons mounted on the right at hip height.

Katja and Al had by now entered the room. They waited for their eyes to adjust to the natural light from the windows. Then they flinched, covering their faces, as the elevator box began to rise, spreading a cloud of dust.

They heard Don cough, and watched the cloud of dust rising with the shadow of the elevator cabin in the open frame.

"Hey, come down," Al shouted, "do you think we want to go up on foot?"

A rattling sound came from above, then a shower of rubble rained down around them. There was a dull thud, a splintering sound, and a crash as a dark object struck the ground in front of them. The floor seemed to tremble and the elevator framework shook from side to side with a grating noise.

Al rubbed the dust from his eyes and made his way through the fog to check that the others had also survived.

"That could have gone badly wrong," he groaned. Next to him, Don emerged from the dust. "Are you still alive?"

"Just a bruised hand," Don muttered between his teeth. "How could I have been so stupid?!"

Al gasped for breath.

"I ask myself the same thing."

"Well, of course, you're the know-it-all," Don hissed, spitting out grains of sand.

"Your obtuseness is getting on my nerves!" Al snapped. "If you don't want to take this seriously, you can find someone else to do it!"

They heard a whimpering sound from the darkness. It was Kat.

"Kat," Don called out softly, "where are you?"

Al also forgot the argument.

"Are you hurt?"

"Yes," Kat whispered.

The two men carried the girl outside and laid her on the ground.

"What's wrong, Kat?" Al asked.

Katja was crying quietly to herself.

Don moved her arms and legs, then lifted her head, and tried to turn her body over. But Katja suddenly jumped up.

"You're getting me dirty," she shouted out angrily. "Leave me alone!"

"Children, children," Al admonished them. "If there's nothing wrong, let's leave it at that! Let's stop arguing!"

Don was offended.

"I'm going to walk up. You can do what you want!" He turned to the door and passed through.

Al glanced doubtfully at Katja – she looked a little the worse for wear, but alive enough, and he hurried after Don. Sighing, Kat followed.

The tower was at least thirty stories high. Out of breath from the steep climb, they finally arrived at the top. Don didn't seem to have been entirely off base with his conjecture. The room was covered with a dome made from the material whose wondrous properties they had already experienced – shiny on the outside, unevenly reflecting the play of colors, transparent from the inside, enhancing the plasticity of the captured image in the most remarkable way. Perhaps they hadn't noticed it before, or the phenomenon was particularly intense here. The stars hung as though distributed through space in cascades.

They stood on a kind of raised platform, their foreheads damp with sweat and dust clinging to their clothes. The shock they had just endured was etched on their faces, but here, for the first time on their journey, they felt the magic of something quite tremendous, something never experienced by humans, which would have roused even the most jaded.

After a short while, Don approached a small box that stood on a pedestal in the center of the room.

"You try it," he said to Al.

Al carefully turned one of the wheels, and they immediately saw the effect it had. The section of the universe displayed on the dome was quite different. A turn of the dial through a mere second of arc had launched them millions of light-years out into space – it was as though they were standing among the stars. The silent lights hung in colorful nebulas across the void, in spirals, disks, and spheres; comets floated between them, clouds of gas turned slowly on themselves. There were unknown star groups, constellations never before observed, but it was still the same universe that they had seen from Earth, the same stars, the same dark or luminous matter clouds, and somewhere, in a hidden corner which they had no leisure to search for now, but which was surely accessible to this wonderful artificial eye, was also the Sun, with its family of planets, the icy world of Neptune, Saturn with its many rings, the deserts of Mars, the cloudy cauldron of Venus, the glowing sphere of Mercury, and in-between, the wandering Earth, the homeland. This was where humans had come into being, where their culture had evolved, where those of whom they themselves were only a part lived and thought, wished and died. None of them would ever have believed that they could still be so amazed, and during those few seconds each felt something inexpressible, which seemed to dissolve the very scale of values they used to gauge what they desired, what they strived for, and what they had achieved, into something absurd and yet meaningful in another, unforeseen way.

Al turned the dial back, and they were surrounded once again by the panorama of the planet, the fortified central peak, the ring-like structure of the outer city, the basin of hills and lakes, and the circle of mountains that enclosed them.

Al touched another lever, and with dizzying rapidity the landscape tipped over and shot towards them. A greatly enlarged section moved steadily along the lower edge of the dome, which was directed slightly outward and downward. Houses, towers, and bridges emerged like a dazzling stage set and collapsed into a point-like nothingness at the edge of the magnified area, sliding later into the insignificant, blurred, and insubstantial, while new buildings ballooned up into sharply drawn backdrops.

"Stop!" yelled Don.

His shout tore them all from the slumbers of this enchanting moment, and they perceived again, where they had only looked before, they heard, where they had only listened before, and they thought, where they had only felt before. And what they perceived was the kind of movement that was by now unexpected for this place, somehow strangely inappropriate. Something typically human – the bobbing up and down of a body while walking, the bending of knees, the stepping forward of feet, the swinging of hands .... Dust clouds rose behind them in little yellow swirls. Every detail was so clear.

"Jak!" Don yelled again. "The one in front is Jak, René is coming up behind him ... and there come Tonio and Heiko, too! Where are they? Can you get their bearings?"

Al zoomed out and they got some idea of how far away from them the second group was, and in which direction they were approaching the center. Then they disappeared behind a long, low building.

"Good heavens, Al!" Don exclaimed. "They didn't get blocked outside either. Jak has the right instinct. They're heading toward the center. How far away from it are they? One kilometer? Two? Al, what are you waiting for? Come on! Katja, hurry up! They'll get ahead of us!"

Al focused on a point on the screen and slowly adjusted the magnification. Now he could see what had caught his attention.

"What's the matter with you, Al?!! Come on!" Don returned once more and tried to drag Al along.

"Just a moment," Al rejoined. "Look at this!"

Everyone looked at the spot that Al had just enlarged, a scene from the modern outer part of the city. In the midst of a group of the familiar teardrop-shaped buildings, surrounded by peaceful green spaces, lay a strangely contrasting spot. Two houses had been pushed to the side and bent in, and a completely crushed transport cylinder lay between them. The center of the destruction seemed to be a flat depression, in which grass had long since grown back.

"What's the matter? It's just another meteorite crater," said Don.

"There are no meteorite craters here." Al closed his eyes, and thought for a moment. "It's much simpler. It was an accident, an explosion." He fell silent again for a few seconds. Then he said: "The outer ring was not the last phase of the city. You were right, Don. We have to go to the center."

*   *   *

Once again, they were on the march toward the city center, but they didn't get far. The houses crowded closer and closer together, and the alleyways they were following began to wind ever more tortuously. The three friends found they could never keep going in the same direction. They were forced into detours, forever turning new corners and entering new squares, without really getting closer to their goal.

"How much longer are we going to keep going?" Katja asked weakly. "It's getting dark!"

"What? Turn back now!" Don was horrified. "You want to turn back now, when we're so close to the goal! And I brought someone like you along with me!"

"I was only asking," Kat defended herself. "There's no need to snap at me. I just wanted to warn you. We don't even have a flashlight!"

"It's going to be a starry night," said Don, brushing all concerns aside with a wave of the hand. "Instead of worrying about that, tell me how we should proceed here. This is such a mess. Tell me how we can make some progress here!"

"I can't help feeling there should be a larger road further to the right," said Al. "That's the best chance of getting inside."

"What are you talking about? You mean going back out then coming back in again? There's no way we're wasting our time like that."

But after they had wandered around in circles for a while, he finally agreed to follow Al's advice. They roamed through the labyrinth for about a quarter of an hour, stumbling over the bumpy, dust-laden ground, past cracked walls and empty window holes, until finally they reached a tall, narrow house leaning against a huge stone wall. The wall was built of roughly cut stone and crudely mortared. Its jagged top seemed to touch the sky. While houses lined their path on the right as before, it was now the wall that accompanied them on the left.

"A city wall," said Katja.

"Now we've reached the Middle Ages," said Al.

Don was still irritated.

"There must be some way to get over it!"

"We could try to get hold of a ladder," Katja suggested.

"Such walls usually have gateways," said Al. "The inhabitants had to get in and out somehow."

Katja slapped her forehead. "Maybe they could fly!"

"Hardly," said Al, "think about the hovercrafts! Why would they have needed those? And everything else we see here speaks against it."

The road in the shadow of the wall was leading slightly uphill, and since the upper edge of the wall ran exactly horizontally, it no longer seemed so insurmountable.

"What about the beings who live here? What do you think they might have looked like?" Kat asked.

"I don't think they would have been very different from us," Al replied.

"Why do you think that? Couldn't they just as well have been giant frogs, ants, or penguins?"

Al laughed amusedly.

"You can tell a lot just from the way things are here. Even though we've only learned a little, it still reveals quite a lot. Remember that the seats are well suited to the shape of our bodies, the control panels are designed to be operated by hands, and the images we've seen on the displays match our eyes perfectly. The sizes of the windows and doors don't differ much from those on Earth. And although there are no steps here, the slopes of the paths are easy enough to walk up, even if we are not used to it."

"You're able to deduce so much," said Kat admiringly.

Don was not taking part in the conversation. His gaze was fixed on the wall, so keen he was to find a way through it.

"But there's another thing," Al continued. "This planet is almost one hundred percent like our own. The gravity is the same, the lengths of day and night match ours, the climate is healthy and spring-like, just like we find it in the spa parks of Ethiopia and Nepal. And I could go on. And this similarity makes it very likely that this world would have produced beings that are broadly similar to us."

"So, you mean," said Kat, "that there might also be …"

But Don interrupted her.

"Enough of the chatter! I want to see what there is behind the wall. Help me up, would you!"

The stone wall was not so high here. It seemed that part of it had collapsed. A mound of rubble bulged against the wall, and this had provided good conditions for vegetation, for it was covered with bushes, and some climbing plants stretched over the stones and up to the top.

Don climbed up the pile of debris, gave the vine a good shake, and began to climb. The plant must have been quite old because the trunk was as thick as an arm, and although some parts were withered, others still looked fresh. Above all, the many forks and branches formed a perfect frame for climbing up the wall. In fact, no one could have imagined a better one.

When Don was up, he let out a cry of disappointment. "Come up!" he called. "We're not through yet, but it's worth the view!"

Al got Katja to climb up first, then climbed up himself. Soon they were all standing on a wide ledge and looking into the interior of the city. It was the sight they had already enjoyed the day before from the helicopter, but now they were much closer to it. They could see the huge castle complex with its wide rusty brown and inky blue roofs, the pointy towers, their flanks patterned with arrow slits, the moist and silvery walls, and the cobbled courtyards. The whole ensemble was crowned by a ruin with a collapsed tower and a wall hanging from it like a broken wing, pierced in several places.

The impression of a child's toy had been banished, to be replaced by one of age and impermanence.

On the other side of the wall stretched a deep moat, overlain by a green-gray mirror of water. On the far side stood another wall, not as high as the one they were now on, but high enough to prevent a swimmer from pulling themself up out of the water.

"Nothing doing," said Don. "Maybe we can make our way along here a bit."

They climbed along the ledge of the wall, went around two slight bends, and reached a platform. It was bordered by a parapet, and a staircase led up to it from the street.

"It would have been much easier here," murmured Al.

Don strode over the stone parapet, and the others followed suit.

It seemed to be a viewing platform. From here, there was a particularly good view over the castle. The wind began to whistle by fiercely. It echoed and howled, filling the air with indistinct sounds. They thought they could make out hammering and clinking noises, muffled calls, vague bursts of shouting carried on the rumbling wind.

Suddenly, Don suppressed a cry. He reached out and grabbed Al's and Katja's shoulders, his fingers digging in painfully.

Something was happening below. It seemed almost like a scene from a dream. Two rows of black-cowled figures emerged from a round gate into one of the courtyards deeply immersed in shadow and positioned themselves on the right and left sides, their hooded faces turned toward each other. They carried blazing torches, which spread an eerie light over the scene. Then a strange rider spat out of the gate, his body covered in gray armor, his head wearing a gray helmet. His mount resembled a large gray weasel. From the other side approached a second rider, a counterpart to the first, in white. Both carried objects that looked very like whips, but much more massive. They held their steeds still for a moment and raised their weapons in salute. Then they charged, swinging at each other, sparks flying with every hit, the sounds reaching the spectators on the wall only a moment later.

"Electric whips," whispered Don.

They watched the events with great excitement. The situation changed rapidly, the two steeds moving like snakes, the whips flying through the air, and the blows cracking. Each fighter began to sway in his saddle at times. Both were showing signs of fatigue, but they turned their animals around again and again, to commence another wild exchange of blows. In the gathering darkness, the sparks streaked like shooting stars through the air. Then the duel was over. The gray knight lay on the ground. The white one raised his whip in salute and rode back through the gate. The rows of hooded figures followed him slowly and clumsily. The torches blazed one last time. Then the place sank into darkness. The display was over.

The three friends looked at each other, Don triumphant, Katja full of excitement, Al in deep thought.

"They're down there!" said Don. "They're still alive and they're playing at knights in armor. They've slipped back into another age! We won't have any problems with them! If only we could get down there! What's the matter, Al?"

Al had broken a loose stone out of the wall. He drew back and hurled it with all his might toward the surface of the water, then watched its fall attentively. But he turned away, disappointed.

"Too dark, pity!"

"Have you noticed something again, Al?" asked Kat.

"Yes," he replied, "but I can't prove it. Not yet. Come on, let's go!"

From the platform, it was clear that the wall was now raised by steps several meters high that they would not be able to climb. But further to the right, beyond another bend, they saw something that made their hearts beat faster: a bridge that spanned the moat in a high arch. They hastened down the stairs.

* * *

The sun had not yet set, and its glow was cast like red-gold velvet over the uppermost parts of the west-facing facades. Below it, the midnight-blue shadow lay like a thick dark liquid. It looked as though it filled the streets, leaking into the walls, intent finally on burying the city under a blanket of gloom. A flaming canopy hung in the sky above. The blue turned into violet, and from the west a crown of yellow-tinged, orange rays began to reach out. Stars were already shining in some places.

Undeterred, Don led the way, and his companions trudged behind. Again, houses leaned against the wall. From their flat roofs, one could have seen over the top, but that was no longer necessary. A new goal seemed to be within reach: to cross the bridge into the center.

Soon, they could no longer see the wall, but they always kept to the left so that they could not miss the access to the bridge.

After a few minutes, they came through a narrow alley, so narrow that they had to walk single file, and then a wide square opened up. Its uneven pavement made it look like a lake whose waves had been suddenly frozen. The paving stones looked strangely transparent, giving the impression that it was somehow bottomless, but this was largely due to the milky glow of starlight, caught in the coating of dust. In the middle of the oval courtyard stood a platform like an island, under a covered ring of columns.

"A well," Kat guessed.

"Or perhaps a pillory," Al suggested.

To the left, the courtyard narrowed, and there gaped the black semicircle of a gate. There was no doubt – the bridge had to be behind it.

They had stopped momentarily and now set off hastily again. The atmosphere seemed to be holding something from them. They walked past the houses with a little trepidation.

"Shh!" Don hissed. He listened for a while. "Didn't you hear anything?"

Katja was about to reply, but at that moment they all heard it – a soft scraping sound. Then silence again.

"It came from ahead," said Al, indicating a rough quarter-circle with his arm.

"It was a long way off," said Don, stepping forward hesitantly. A hushed call from Al made him stop again.

"Someone's been here."

Al pointed to a long thin mark that ran through the dust towards the gate. Don squatted down in front of it to examine the details.

Katja had taken a few steps back and was leaning against a wall projection, hunched over.

"The knights …" she stammered, "the knights with their terrible whips!"

"Don't worry! What can they do to us?" Al reassured her. "You don't need to be afraid. You just need to think …"

Don stood up suddenly.

"It's much worse," he said in quiet anger. "It's Jak and his group." Then, turning to Al, "Look at this trail!"

Al examined the spot and found prints of crepe soles that clearly confirmed Don's observation.

"They got here before us!" Don whispered. "They beat us to it!"

Al joined him. Now he almost felt sorry for his friend.

"Don't lose hope so quickly. Sure, they got here before us. But that doesn't mean they've completed their task. The whole thing is still up in the air."

"Do you really believe that? Are you just saying that to comfort me?" It was clear that Don was regaining courage. "Do you really think there will still be difficulties once we reach the center?"

"That's very likely when the real difficulties will start," said Al. He didn't consider this a particularly positive thing, but Don took it as such.

"Let's get on then, they can't be that far ahead!"

Katja was significantly less afraid of Jak, René, Tonio, and Heiko than of the mysterious city dwellers and breathed a sigh of relief. She quickly joined her companions who were now following the trail. But for safety, she hung a little behind them.

Soon they were passing through the gate. It looked exactly like one would imagine a medieval city gate. Next to the wide passage for wagons, draft animals, and mounts, a narrow pedestrian gate led through the old walls, separated by thick pillars. The footprints ran straight through the middle.

Beyond the gateway, there was a level open space, much smaller than the first courtyard, with two rows of stone benches. It was separated from the water by a parapet on which several figures were standing, apparently looking down on them.

Don approached one of them.

"They're all hooded," he said, and returned to Al and Katja.

"Careful! Don't destroy the tracks!" Al warned.

"What do you want with the tracks?" Don asked in his usual arrogant tone. "They lead to the bridge. Obviously!"

Al was not easily swayed once he had made a decision. He calmly continued his examination of the footprints, to Don's increasing annoyance. The lighting conditions were better here than between the houses and gave him a better chance of getting results.

"You always know better," Al muttered under his breath, not caring whether Don even heard him, "but you don't know everything. There's a tangle of tracks here, leading in all directions. What could that mean?"

Don was already on the bridge.

"They must have stood around here for a while, just like us."

"But we have a reason for it," Al continued, "they didn't – or else they had a completely different one. It would be interesting to find out what exactly they were up to here."

"Interesting," said Don. "Interesting!"

Meanwhile, Katja had walked over to the western parapet, which could only just be made out as a silhouette. She stood below a stone figure, and it seemed to her as if it could step down from its pedestal at any moment and enact some cruel deed – as a judge, an executioner, or a torturer. She forced her gaze away from the silent figure and turned to admire the last glow of the sun which had just sunk below the horizon. Its light was still filtering through from behind the silhouettes of the houses. The view there was clear. She could see the castle moat running straight west, and in its waters, the image of the city, cast for the second time as a dimly shimmering, upside-down mirror world – the converging ramparts and the black tiered structure of the houses on the right and left, and the orange-yellow of the sky with its blood red and dirty brown contour.

Katja suddenly felt strangely alone, vulnerable, and threatened. She saw her companions crawling around on the ground and gesturing to each other, but she did not understand what they were trying to do. She saw the bridge stretching across the water into the night, and on the other side the terraced roofs of the fortifications, their highest, tower-like structure right above the gate, the hundreds of black window openings, the strange devices standing on the flat rooftops – perhaps ancient but nevertheless terrible weapons, just waiting to spew death and destruction again. Eyes stared down at her from every corner. At every window, faces were twisted into silent laughter and fists were clenched threateningly.

"Why did you actually come here?" a voice inside her asked, "what were you expecting to do here?"

Don was also filled with hopes, fears, anger, impatience, ambition, reck-lessness, and sometimes even reasons. So close to his goal, he felt held back, betrayed by his companion's pedantry. He would have preferred to set off alone to get the fullness of the experience, daring some mind-blowing adventure, even if it meant facing violent death.

Like Don and Katja, Al also had something to sort out in his mind. His imagination worked like a chess player's, always trying to read his opponents' intentions from the way they moved their pieces. He was working out dozens of possible combinations by considering the steps his opponents might have taken here, trying to imagine their movements and goals. But none of his mental images seemed to be making sense.

"The track on the bridge is doubled," he said, shaking his head. "That means they came back." He could not yet grasp the consequences of this realization, but he sensed its significance. "They must have come back."

Suddenly Katja cried out: "Stop arguing! Let's get out of here! Why are we hanging around?" Her voice sounded strained and shrill. She ran out onto the bridge. "Let's get this thing done!"

Don shot after her, with Al close behind, his calculations forgotten.

Katja was by now only recognizable as a fleeting shadow, the darkness concealing all details, so Al did not immediately understand what was happening. Only a splintering crash in front of him and a clap of thunder behind forced him to believe what he had glimpsed in the flash of light: a dark, irregularly shaped hole had been torn into the bridge, right in front of the running Katja. A shower of stone rose up and, carried along by her momentum, the girl's collapsing body slid towards the opening and disappeared. There were three more loud crashes, but only one projectile grazed the bridge, making it tremble and swing, and almost lifting Al off the ground. And this time he recognized where the shots had come from. They had been fired from the battlements of the tower behind them. He struggled for a moment to get his balance and rushed forward, out of the range of the deadly hail.

Don stood motionless on the bridge, a few yards from the impact site. Al dragged him away as quickly as he could. Only a narrow strip of the bridge was still intact, and even that was riddled with cracks. They felt it shifting when they passed the bottleneck as quickly and carefully as possible.

Al stopped and peered at something down by the water's edge.

"Kat...", he gasped, "we have to ..."

Don pushed him forward roughly, breathing heavily. "Leave her ... serves her right!"

They ran on, hunched over, close to the parapet so that they would at least be protected on one side. They were not yet out of the line of fire. Something hissed past, then something grazed Al's head, and another object bounced off Don. A long object crashed onto the stone floor. It was an arrow.

Once again, a hail of projectiles rained down on them. And now a horrifying chuckle echoed through the night.

Don stopped abruptly. He recognized that voice.

"Jak," he whispered.

"Who else?" Al asked.

"Hey, Al!" Don's tone of voice had changed from one end of the scale of emotions to the other. "That's Jak and his men!" He was almost jubilant. "Al, my friend, do you see what this means?" He didn't wait for an answer. "Jak is behind us! We are the first! We'll reach the center first! We're the winners!"

* * *

They had covered exactly eighteen meters on the bridge, and those eighteen meters had taken them a hundred seconds. With those eighteen meters, they seemed to have come closer to their goal than on the long journey from the mountain slopes, across the meadows, and through the city to the gate. And in this short interval of one hundred seconds, more had happened than they had experienced during the entire two days of their stay. Three times now their mood had swung to extremes – first to a kind of hopeful determination when they had felt their goal slipping away from them, then to terror, almost burying all their hopes, and now to a reckless confidence in victory.

They stormed on as another shower of arrows rained down, but it was as ineffective as the first. Jak's booming laughter echoed again, but they barely paid attention, because in front of them rose a black, starless block against the equally black star-studded night sky, the center of the city, a castle with countless outbuildings, a fortress ingeniously protected by water and stone, yet now vulnerable to their assault. No light shone from it, no sound was heard, not even a breath of wind – like gigantic sleeping animals, the buildings stood within their ring of water.

Al and Don simply ran for it, and everything happened indescribably quickly. Al just managed to grab Don by the shoulder, and he had to throw himself back with all his might – because here the bridge unexpectedly ended, hanging freely in the air as if cut off by a huge knife, cleanly separated from the other part, of which nothing could be seen anymore, shaved off, without regard for mechanical stability or static equilibrium.

The two men couldn't believe their eyes. Perhaps the darkness was deceiving them? Perhaps their own excitement had somehow tricked them. They reached out with their hands and they felt the sharp edge, a smooth, vertically dropping surface. They peered forward, and saw nothing in front of them but emptiness. Somewhere far beyond, much further than they had expected, there rose something massive, its edges flickering uncertainly in the starlight, a dull glow flitting over metal. They got down on all fours and crawled forward, but they were staring into an abyss so deep that it seemed to take seconds for their gaze to reach into the swirling and weaving, spinning and drifting darkness at the bottom.

"It's over now," Don whispered. "We're trapped."

"Now we know why they turned back," Al said.

"They tricked us, those scoundrels!"

"I'd so much like to know how to get into this damned city."

Don exhaled sharply.

"I don't care about that now. We've lost. Jak can do whatever he wants with us. I think we just have to give up."

They made their way slowly and carefully back over the bridge, past the shattered site of the first hit, and stumbled over the rubble left by the second impact, a glancing blow that had torn away part of the railing.

Don cupped his hands into a megaphone and shouted: "Jak, you've won. We surrender!" And again: "Jak, do you hear? We give up!"

Then Jak's voice sounded: "Hello, Don! Glad you've realized it! Come closer, but slowly!"

Step by step, they approached the gate. Suddenly, there was another flash. It struck next to them. Then another. They were shaken by a crunching sound and Al felt a sharp blow to his right hand. He ran his left hand down his right arm and was startled to find a gaping cut in his hand.

Don received a more serious blow. A splinter hit his chest and another his hip.

"Jak, you son of a bitch," he roared, "you son of a bitch …" Still roaring with rage and pain, he stormed across the open space in front of the bridge. Running to the left, where he had seen the flash in a window, towards the shots, he jumped high, stretching himself up. His hands grasped the lower edge of the window and he pulled himself up. For a moment his face stared into the opening, but then came a flash right in front of him. His body fell over the sloping wall and hit the ground.

Al too had run through the hail of bullets, but he had headed towards the gate and reached it. Here, he was safe from the attacks. His heart was beating fast, and an absurd joy filled him. He felt a kind of atavistic pleasure in this senseless struggle. Pressed against the wall of the gate, he wondered whether he dared to venture across the open space.

A scraping sound came from behind one of the pillars. Al froze. Someone whispered almost inaudibly:

"Al, is that you?"

Al leapt up and, reaching around the pillar with his uninjured left hand, got hold of a shadowy figure. He lifted his foot to kick hard at his opponent when suddenly he heard:

"Stop, Al. Stop!"

He was prepared for a ruse, but hesitated for a moment.

"It's me, René. Listen, Al, I want to help you!"

Al grabbed him by the throat, leaving him gasping for breath.

"All right. Jak is going too far. I'm not with him anymore. I can help you."

"How are you going to help me?" Al asked.

"There's a side door here. Come with me and I'll show you!" Still prepared for betrayal, Al crept after the other. They went steeply downwards, diagonally through a cellar into which a little light had managed to seep from the outside, and finally up another slope. René now stopped and opened a sliding door. It made a grating noise as it rolled aside.

"Keep quiet!" Al whispered.

They listened. Nothing stirred. They squeezed through a narrow gap and looked around them. They were in the large square in front of the gate. They looked up at the battlements where the guns were stationed. Still nothing stirred.

In the shelter of the wall, they began to run, first a few meters, then stopped again. They peered around …

There came a flash from above, from somewhere next to the tower construction, above the cornice. Directly in front of them, something bright and round shot down and shattered. They had no time to think. They were already torn to shreds.

The Second Attempt

"So here we are," said Don.

"Nothing has changed," Katja noted.

"What should have changed?" Al asked.

In the morning, they had set off from their campsite, and now, at noon, they had reached the wall. René had come with them. They were standing once again at the spot where the stairs led up to the lookout.

"Do you think it'll work here?" he asked.

"As well as anywhere else," Al replied. "I don't believe there's any obstacle at all."

Katja was amazed.

"But what about the bridge!"

"I've been thinking quite a lot about that," said Al. "The bridge ends. That's true. But simply because such an old bridge has no place there."

"I don't understand," grumbled Don.

"Well, let's give it a try," Al suggested. "Come with me!"

He climbed the stairs with a bundle he had brought from the camp tucked under his arm. Before the last step, he waited for his companions and examined the stone parapet closely. Finally, he nodded in satisfaction. He pointed to a spot on the right: a small but shiny gray button was embedded in the stone. Al moved to the left, where a similar button was located at the same height.

"Now watch!" he said.

He leaned against the railing and looked down. This time the lighting was slightly different than in the evening; the sun's rays fell almost vertically from above, dazzling them as it reflected off the rooftops and making the depths of

H. W. Franke, *The Orchid Cage*, Science and Fiction, https://doi.org/10.1007/978-3-031-60499-7_2

the alleys and courtyards look all the darker. The angle of incidence was low enough for the light to catch on all the upper features: on the overhanging roof edges, on the protruding cornices, on the rows of battlements, on the bird's nest-like bay windows and balconies, on all the extensions and decorations that covered every shred of free space for no apparent reason. This broke the clear outlines and made the buildings distinctly less imposing.

The courtyard where the mysterious duel had taken place lay before them like a dark pit. As before, a strange mood came over them. The air was full of indistinct noises. There was a rushing, echoing, whistling, hammering, and clinking, and dull cries rang out, the wind bringing with it rumbling sounds and fragments of shouts.

And then the black shapes appeared with their torches now sparkling like weary will-o'-the-wisps, followed by the two riders.

Everything unfolded just as they had experienced it once before: the whip duel, the victory of the white knight, his silent greeting, and the withdrawal of the hooded figures.

Then everything lay still and deserted as before; there was no living creature to be seen, no breath stirred the air.

René, who was seeing the event for the first time, needed some time to recover from his astonishment. Katja whispered: "I don't understand," and Don exclaimed: "Just like the first time! Not a single detail was different! Al, is it some kind of show?"

"Something like that," said Al. "It's like a film. The perfect illusion produced by technical means. The audience enters the stage and a photocell registers their arrival," he pointed to the two buttons in the railing. "It passes on the message and the show begins."

"How did you come up with that?" Don asked.

"I just remembered that I once saw something similar on an old television recording. Some moving figures were connected to a clock. At noon, they moved out on a rail and performed a little dance, with strangely lifeless, puppet-like movements. Then they disappeared again. Here, the illusion is much better done, of course. But it was the way they entered the scene at exactly the moment we stepped onto the ramp! That made me think of those puppets and I had the explanation."

"But why didn't you say anything about it?" Don asked suspiciously.

"Because I couldn't have proven it."

They looked back across the moat to where the hill with the ruin rose.

"What actually is real about all this?" René asked.

"Nothing," said Al. "I don't think there's anything real there." The others looked at him, shaking their heads. He reached into his pocket and pulled out some pebbles. "Watch closely!"

He threw a stone. It fell in a wide arc, almost placidly following its parabolic path and disappearing under the water's surface. But René felt that something was wrong. He didn't know what. The stone had flown through the air and was gone – something was missing, right? It hadn't hit anything. No water had sprayed up, no waves had circled out from a point of impact.

"There's no water," said René. "The water is part of the game."

"... and the houses, the streets, the hill?" Al asked.

"A backdrop," said René.

Don squeezed himself between Katja and Al. There was a frantic look in his eyes.

"But what's behind it?"

"I don't know," Al replied. He opened the carrier bag he had brought with him and took out a ladder rolled up into a tight roll. The rungs were made of light steel and the cords of duralumin wire. He also pulled out a short piece of braided wire with rings attached to its ends and wrapped this three times around the horizontal upper part of the parapet. He then clamped the rings into carabiners at the end of the ladder. This secured one end of it at the top and he let the other fall down. The bundle unrolled as it fell and it plunged unhindered into the water. The ladder swung back and forth a while, then hung still.

"I'll go first," said Don, looking around questioningly. Since no one objected, he climbed over the parapet and made his way down. Rung by rung, he went deeper until finally he touched the surface of the water, but there was no reflection in it. He dived in, but felt no moisture. He dived under, but found he could breathe. He had intended to shout, but just looked around speechless.

The two men climbed after him, first Al and then René. Katja watched all three as the shimmering darkness below swallowed them up. All this happened in a chilling silence. The surface did not so much as quiver. The ladder trembled a little and then was still. When the last of her companions had disappeared, she suddenly felt indescribably lonely. She started to climb over the parapet, but hesitated. She tried, but couldn't actually manage to do it. She stared across at the castle, but she wasn't looking at the walls and towers. She was trying to see through them. Concentrating as hard as she could, she tried to penetrate the veil, to see behind the curtain, but nothing came to her except the terrifying images produced by her own imagination. She was aware of this and yet those images tormented her.

Then a voice called up from below, just a few words that she couldn't quite grasp. Her name, and perhaps something comforting. Suddenly, she found she could move again. She climbed carefully down and experienced what the others had experienced a few moments before. She was hanging in the water, or rather, in what looked like water from the outside. Her eyes lowered to the level of the surface and then the picture she saw seemed to flicker and distort. Something flipped, and the previously wavering image came together into a solid form.

Now there was no sign of water. She stood with the others on a softly flashing metal surface, extending directly away from the base of the wall. Where once there had been a scene from the Middle Ages, there were now airy buildings with glass roofs supported by pencil-thin pillars, with walls of stretched wire mesh and long parallel strands of coiled pipes. Antennas and parabolic mirrors were raised high on what looked like long rods, while all around lay objects made from metal and plastic and glass, for which they had no names.

This was the real center of the city. A monstrous, flashing body of machines.

* * *

Al leaned with his back against the wall, as if he wanted to maintain the last connection with the normal world for as long as possible. Don tried to find something familiar or at least something that he could explain among the structures in front of him, a recognizable feature that could help him regain some kind of equilibrium. Katja was looking for a place to sit down, but she searched in vain – no human needs had been taken into account by those who had built this open space. René scraped the ground with his foot, then squatted down and knocked with his knuckles on the solid, gray mass as if to test it, before standing up and looking around patiently.

"Looks quite different now," Don noted. "There's no trace of the old town, or the illusion of it." He looked at the ladder, the only connection with the outside. It hung there, stretched out and smooth against the wall. There was no distortion, and nothing to show that it passed through a zone that was optically activated in some inexplicable way.

"I don't like it here," Kat grumbled, "it's so …" She searched for a suitable word, but found none.

"Uncomfortable," said Al in a slightly mocking tone. Katja thought hard.

"Different," she said. "Alien."

"We must go on," Don urged.

"Where do you want to go?" René asked.

"Now listen to me!" said Al in a louder and more decisive way than was his custom. "We've decided to try again. Fine. Now we've got where we left off three days ago. But we mustn't believe that it'll always go on as before; that we just need to wander around as though we're in a nature reserve and that what we're looking for will simply show up of its own accord. This could get dangerous. We can't just ignore that possibility! Here there's …"

"So, you do think they're still alive?" Katja interrupted, retreating discreetly back toward the ladder.

"I think they've reached the end of their journey. The eerie thing about it is this: we don't know how they evolved after that, when the pleasant stage of doing nothing in their garden houses came to an end. We also don't know how we'll continue to develop ourselves. And that's why we're likely to encounter things here that we've never heard of before, machines whose behavior we can't predict …"

"How should a machine behave?" Don asked. "You press a button and it does what it's set up to do."

"It can also be more complicated," René said. "You may not even need to press any buttons. It may just do what it needs to do on its own."

"… what its program prescribes," Al corrected. "But what happens if it sets up its program itself?"

The question hung in the air for some time. Katja couldn't imagine what would happen then, and she wasn't interested either. She wondered if she wouldn't have been better off at home, and began thinking about the adventure films that were never as exhausting as this trip, and where there was never so much discussion; where heroes fought battles in single-occupant spacecraft and she then fell into the arms of the victor; where she danced with the idols of the classical era, with Fred Astaire and Frank Sinatra; where she was Cleopatra, ruling, condemning, and seducing, and where Caesar and Augustus lay at her feet. She thought about the remote-controlled game boxes connected to every desk, about the rolling, bouncing, and floating balls, and about the ringing of strikes and the sharp bang as they burst when points were lost. She thought about the ballet of colors and shapes, about floating in three-dimensional spaces, and about the mixtures of tastes and scents. But strangely, she couldn't get excited about it. It's actually quite boring, she thought. Maybe something else will happen here after all. Sleepily, she leaned back and closed her eyes.

What happens if machines set up their own programs themselves?

Don had imagination. In his mind's eye, machines began to swell up, sprout, and proliferate; lines split, pillars bent, walls bulged, and a mad chaos of spokes, wheels, T-beams, pistons, tubes, ball chains, wires, transistors, ther-mocouples, magnets, rubidium crystals, relays, potentiometers made of glass fibers, polyester resins, viscose wool, rubber, slag, and gelatin sprang to life. Metal claws groped around like the shoots of degenerate plants. A circuit with its own will shocked its victim like a kind of sophisticated torture device. A blistered mass swelled up and burst apart, shooting out sticky tentacles like a polyp. Insane robots threw themselves at helpless people tied to their chairs, while entire armies of them broke through the low-rise buildings of peace-loving settlements, driving a wedge of hatred and destruction.

In a strange way, these ideas excited Don. They generated disgust and fear, but they also activated his readiness to defend himself, to rebel, to wreak revenge.

Suddenly, the image had gone, and before him was this clean space, filled with the incomprehensible, but undeniable order of an alien technology. He twisted his lips in contempt and turned back to his companions.

What happens if they set up the program themselves?

René had a special relationship with machines. He understood them in the way others understand a piece of music. He knew so much about them; about the meshing of cogwheels, about the interactions of switching elements, about the forces in materials, air, and vacuum. And where this understanding ended, there began the conviction of a meaningfulness behind the thousands of movements and impulses, the causes and the effects, the circling, flowing, and vibrating, the actions and their consequences. The machine that set its own goal had become for him a symbol of all that is functional, a symbol of all that has meaning, from which all arbitrariness has been removed – *l'art pour l'art* at the highest level, that can no longer be surpassed. Was that the case for these structures before him? He did not agree with Al. They might well have been the product of some remarkable intelligence, but they were not machines that existed for themselves. Why? Because they accomplished no function. He saw no movement in them and felt none of that special fluid that emanates from current-carrying conductors, from pulsating electrons, from oscillating fields, and the like.

At the end of his considerations stood disappointment.

"You're making me quite sad with all your talk," said Don. "What do you actually want? Do you think someone will attack us here?"

Al wanted to answer. He looked at Don, then at Katja, then at René. They didn't understand what was going on. He remained silent.

"That's what matters," said Don. "We have to be realistic! We can't afford to die again. Jak is probably already here. He is three days ahead of us. If he's not already at the destination, then we've been lucky. Jak is the biggest threat. And now, Al, speak up. What does all this magic mean?"

"I believe the old town, along with all the figures and their hustle and bustle, is just an attraction. There are probably other viewing platforms along the wall from which similar spectacles can be observed. In reality, the machines that produce all that are located here in the center, but they had other tasks as well – namely to produce energy for the inhabitants, to provide them with food, comfort, and pleasure, not much different than what machines do for us today."

"I have to ask again," said Don impatiently. "How might the machinery become dangerous to us?"

"How should I know?" Al returned, slightly annoyed. "I've told you what I know – now draw your own conclusions!"

René took a step forward.

"What dangers could there be? The facility is built for intelligent beings – the more perfect it is, the more it will try to satisfy their desires."

"I can't say I see any of that," said Kat, standing up. "There aren't even any benches here. If you were right, then there would at least be a taxi to pick us up. This traipsing around is getting on my nerves." She took a few steps away from the wall, across the horizontal ring on which they had arrived and into a wide street lined with a grid-like structure.

She crossed one of the lines …

"Look!" cried René.

A slate-gray structure came gliding towards Kat, its perforated prow stopping just in front of her. Then it turned and presented its flank. The outer surface split into two parts that moved apart and they could see into a glass-covered hollow, similar to the interior of a boat, with thickly padded seating all around. A board slid out and unfolded, neatly bridging the gap between the step and the street floor.

"How nice!" cried Kat. "Just what I wished for!" And with one step, she was in. "Come on!"

"Wait, be careful!" cried Al, but Don just laughed and climbed in. When René followed without hesitation, Al also got into the vehicle. Don had gone forward so that he could look through the glass at the front, in the direction of travel. "Where are the controls?" he asked.

But there were no controls. There was nothing that even remotely resembled a control.

"So, how do we get out of here?" Don asked.

The ramp slid back in place and the sliding door closed. The vehicle started moving.

"Stop!" cried Don. "Where are we going?" He looked for a brake, but there was no brake. He looked for a door handle, a knob, a lock, anything, but there was only the smooth wall, the cushions, and glass. "I'm afraid we're trapped," said Al.

* * *

Through the curved glass walls, things flashed past that they had already seen from a distance without being able to make any sense of them.

Don kicked angrily against the door, but all he achieved was a sore foot.

"How do we get out?" asked René. "There has to be some way out."

Al was gazing at Katja, who was rocking happily on the springy seats. He felt a trace of envy – the envy of someone who is always fretting when they encounter that blissful state of not caring in someone else.

"How do we get out of here?" he repeated. "Quite simply! We just need to say or think the right thing."

Don turned around irritably: "And how do we do that?"

"Good question," said Al.

René went to the front and loudly gave the order: "Stop! Stand still! Stop!" After a while, noting that they were still pursuing their hovering journey at the same speed, he added apologetically: "Well, it might have worked."

"What now?" asked Don.

"Wait," advised Al.

They were passing through a network of metal structures where fine, decorative webs of white threads formed ornaments inside black frames, glass surfaces flashed coldly in the sunlight, and reflections of light picked out the shadows on the gray ground. Whenever the vehicle turned a corner, everything inside rotated as if on a revolving stage. But suddenly, it reduced its speed, stopped, and moved to the right at right-angles to the original direction of travel until it came to rest against the wall of a large block-like building.

The door gaped open, and a door of the same width also opened in the wall.

"Get out!" René shouted.

But Katja now sat motionless, as if frozen.

"Not so fast," shouted Don. "Who says I want to get out?"

René got up calmly from the bench and stepped through the opening. There was a click, a bright horizontal line dropped from top to bottom and René was swallowed up by the half-light. "Well," Don said, and followed his example. The line appeared again, accompanied by the clicking sound.

Al peered into the adjoining room, which was dimly lit, but he couldn't say where the light came from.

"Hello, René! Hello, Don!"

He listened, but got no answer. He called again: "Hello, Don!"

Nothing.

He felt the delicate touch of fingers on his neck and turned around. Katja stood in front of him. Her eyes looked unusually dark. She grabbed him more firmly, her hands clutching his shoulders, pulling him away from the eerie portal into the unknown. She pressed herself against him, seeking protection, keen to drown out the eeriness with something more immediate, obsessed with a desire for human touch, for some kind of anesthetic, something to help her forget – even if it was only for a few seconds. She pressed herself against him, closed her eyes, kissed and let herself be kissed. She saw and heard, hoped and feared nothing more, because she didn't want to see or hear, hope or fear anything. She gave herself up expectantly to the pleasant, calming, and at the same time exciting sensations, and tried to escape from reality, from the vibrations, electrons, atoms, metal and plastic, circuits, images, and intentions, and she succeeded as perfectly as she could have wished for. She was swept into a whirlpool of dizzying feelings, but consciously kept alive a tiny remnant of reality during these moments. In some dark corner of her mind, she savored to the last drop everything that was strange, absurd, even contradictory in the situation.

When some of the intoxicating confusion had subsided and he remembered the urgency of the present, Al noticed almost astonished that nothing had actually changed. No one was trying to force them to pass through the door. No one seemed to mind that they had stayed behind, and it seemed that they could remain apart as long as they wanted. On the other hand, the door remained invitingly open and the vehicle was still in exactly the same place so, strictly speaking, that was in fact a kind of coercion: even more inescapable than any act of violence.

"We have no choice," Al said softly. He put an arm around Katja's shoulder and stepped with her into the portal and through it.

Just as they crossed the threshold, a bright line dropped down between them like a gentle flash of lightning. It was as if a wall was being pulled down. They were separated.

Al stood in a gray cubicle. There was a flash of blinding light, then darkness fell over him like a black cloth and the floor began to carry him towards the wall. Just as he reached it, the wall split apart and then immediately closed behind him. He now stood in another cubicle. The right-hand side was covered with a grid, from behind which there came a soft rustling sound.

There was a faint bang and the rustling sound was abruptly cut off. The floor moved, the wall made way for him and closed again. Something whirled up from the floor and rose over him. Now there was a faint smell of chemicals in the air. The floor shifted, the wall slid apart. And so it went.

The first few stressful seconds took his breath away and he was incapable of any rational thought, but now the whole experience suddenly changed. It was no longer something abstract. Rather, he was overtaken sharply by a sober realization: he was being cut open, disassembled, broken down in some non-mechanical way. And this cold certainty was more crushing than trying to resist something undefined. The skin on his hands had gone numb and his tongue lay like a rubber ball in his mouth. Suddenly, he thought of Katja and forgot everything around him as he cried out: "Katja, can you hear me?"

"Yes, Al, I can hear you."

"You don't need to be afraid."

"Of course not, Al."

"Now only one thing is important. You, Katja!"

"And you, Al!"

The floor slid forward once more and the wall tore itself apart to reveal an empty cubicle. The right-hand wall was covered with a honeycomb pattern of circular holes, and from one of them situated right in the middle, a blunt-tipped arrow began to move towards Al. He pressed himself against the front wall so that the arrow passed behind him, stopping in a horizontal position on the left-side of the room. He had escaped! But no, not yet! A second blunt arrow sprang out of the wall at knee height, horizontally like the first. Al dodged and the arrow passed. But it was immediately followed by a third, chest high. He ducked. The narrow room was already constricted by two arrows, and now there was a third. A fourth soon pushed its way into the cabin, neither slowly, nor quickly, with an automaton-like regularity. It stabbed into the narrow space, again at chest height, directly towards Al's trapped body. He had trouble ducking this time. The rods now seriously hampered his movements. Another went over his head and now another was coming towards him. Al crouched on the floor, trapped in the ever-narrowing grid. He tried to dodge, pulled and shook the bars, but it was no use. This time there was no escape. He turned his back to the approaching spear and waited. He felt a dull pressure below the shoulder blade, then a snapping action and something like a thorn that drilled into his skin, causing him a sharp pain.

As if on cue, all the rods retracted. A few seconds later, the room was empty again, leaving only the honeycomb pattern on the right to remind him of the torture.

The floor moved. The wall gave way and closed. A nozzle protruded from the right into the room and began to hiss.

...

"Katja, answer!"

"I'm answering, Al."

"Please don't leave me alone!"

"No, no, Al."

"Are you happy?"

"Yes, very! I just need to think of you."

...

He was being transported by a conveyor belt. At each station, something different happened to him – unusual, frightening things, not so much painful as tormenting because he had no idea of their purpose.

Station ...

A light began to glow, faintly at first, then gradually intensifying, until it became all but overwhelming. Al pressed his hands to his eyes, and yet still he seemed to bathe in this blinding flood of light.

Station ...

It began to get warmer, slowly at first, then faster, until the air was scorching hot. His skin burned, his heart beat, his lungs gasped for breath. Al writhed, gasped, beat his fists against the walls.

Station ...

There came a humming sound, softly at first, barely audible, but becoming fuller, until it filled the room, loud, powerful, booming, thundering, roaring. Al hunched his shoulders and crouched on the floor, holding his aching skull between his hands.

...

"Katja, I couldn't bear it if you didn't ..."

"Stay calm, Al. Please! For my sake."

"I am calm, Kat. Where are you now?"

"I'm not paying attention anymore. Why should I?"

...

Why indeed?

There was a lens looking down into the cubicle. It was the eye of the machine. The wall to the right fell back to reveal an abyss and an eight-foot-long creature began to wriggle its way across the floor. A swing-wing plane buzzed towards him, teeth bit, faces grimaced.

Why pay attention? Why?

Al reached into a face, then reached through it.

The conveyor belt started to move again. There was a new cubicle, empty except for a red button on the wall.

A tingling feeling ran over Al's skin, growing stronger, then weaker, then stronger again, then much stronger. Desperately, he looked around for a chance to escape? Was there anything that could save him? He found the red button and pressed it. The electric tingling feeling stopped abruptly.

The floor began to carry him away and the tingling sensation ran through him again. Al searched for the button and found it, but it was not fixed in a holder. It was movable. In fact, a tangle of lines was cut into the wall in the form of a labyrinth, and the end of one downward-running incision was circled in red. The vibration became more intense, then subsided, then surged. Al was already moving the red button. Only twice did he get lost in a dead end and have to turn back before he found the way through the maze and could press the button at the indicated place. The electric shocks subsided immediately.

The belt took him further …

There were other tasks. His limbs were trembling from the electroshocks. He was thinking hard, with all the concentration he could muster. Al took it as a challenge, as a test of endurance. He did his best and was proud when he succeeded.

…

"Leave these things, Al!"

…

"Al, what's the point?!"

…

"Have you forgotten me so quickly?"

…

"Give up, Al! If you love me, give up!"

…

Al was putting building blocks together to form a cube, looking for matching parts from a pile of metal plates, reacting to the glow of shining disks, solving simple and difficult arithmetic problems, and still the conveyor belt kept moving. Then the wall opened up, bright sunlight blinded him, and he staggered outside.

There sat Don, René, and Katja. They looked a bit exhausted, but otherwise seemed fine.

"Well, survived?" Don asked.

"What took you so long?" René said.

Al looked at Katja.

She sat against the wall with her knees drawn up and stared past him. Her lips were curled in contempt. She was whistling to herself. Al took some time to collect himself.

"Where are we?" he asked.

It was René who answered: "At the back of the building."

"Where exactly?"

No one knew.

Al approached a structure like a truss, perhaps a drilling tower, and climbed up. The physical exertion drove away the fatigue and the remnants of the horrors he had just endured. It was like a refreshing bath. He quickly climbed higher, above the level of the rooftops.

There was a current of warm air which cooled him down pleasantly. His companions had become small, inconspicuous dots. He surveyed his surroundings. The hovercraft had brought them to the northern part of the city center. Between the taller buildings, there were enough gaps for him to see the city wall, which curved like a bowl around the enclosed horizontal surface. The buildings with their metal and glass roofs were nestled in this basin like the components of a neatly arranged construction kit. The flat area was disturbed only at one point – Al assumed it was the same place where the hill with the ruin had been in the picture of the old city. Here the buildings rose higher. Al could not tell whether they were standing on a hill or were just taller than the others.

He climbed back down and reported to his companions.

"I have an idea," Don announced as they discussed their next steps. "Jak is three days ahead of us. Let's try to find him and his people, then we'll see what he's up to and we can save ourselves the trouble of searching ourselves."

"Good idea," said Katja.

Don turned to Al. "Did you see any sign of Jak?"

Al shook his head.

"No."

"No matter," Don stated. "It's not a large area so we'll soon catch up with him. The hill seems the most interesting place to me. We should go there first. But we'll need to be careful, because Jak will guess that we are back by now!"

"Will we be able to move freely?" René asked.

"Why not?" Don replied. "The machines have tested us, that's clear enough. And they've let us go. They've judged us harmless. They won't care about us anymore."

Al disagreed once again.

"I don't believe they don't care about us anymore." He pointed to a pillar standing in the middle of the nearest large square. There were many of them: thin stalk-like structures that ended in dark, shiny spheres of indeterminate color. Some of them were only a few meters high, others towered over the rooftops by many times the height of a man.

"Lamps?" asked René.

"Maybe that too," replied Al, "but I reckon they're also eyes."

René nodded.

"Ball lenses."

"Eyes staring at us all the time," said Katja, without making it clear whether it was a question or a statement. "Thousands of eyes watching us constantly."

"That's just a guess," said Don uneasily.

"We have to take such suspicions seriously," said René. "We can't just dismiss them!"

"Well, go after them then!" Don challenged, rudely.

René replied coolly: "That's exactly what I intend to do."

He strolled over to one of the columns and took off his jacket. He tied the two sleeves together so that they formed a loop, and hung the garment over his left arm.

"Now you'll see you're not the only one that can climb," he announced loudly to Al, who was following behind slowly with Don and Katja.

René grasped the pole as high as he could, pulled up his legs, wrapped them tightly around the smooth plastic shaft, stretched himself up again and in this way climbed up surprisingly quickly. After a few such movements, his head was close to the sphere, and he jerked back instinctively. Although nothing moved, it seemed to him that the gaze of the round glass body was decidedly evil. Skillfully, he took the jacket off his arm and with a swift movement pulled it over the glass eyeball. A slightly oppressive feeling came over him and he quickly slid down and joined the others as if he were trying to hide among them.

Despite the undeniable harmlessness of their actions, they did not feel entirely comfortable about it. They looked around uneasily.

"Child's play," Don muttered dismissively, but in reality, he was just trying to reassure himself.

Suddenly, something came buzzing over the rooftops and hovered in front of the covered sphere – a kind of metallic bird the size of a condor. A claw reached out and pulled the jacket off the sphere. It abruptly set itself in motion again and buzzed towards René, who took a step back, startled. When the jacket fell to the ground, the flying object was already gliding away over the rooftops.

"You may as well put it back on," said Al. Dazed, René picked up the jacket but had to make several attempts before he managed to get his arms back into the sleeves.

"So now we know," said Don. "Well, so what? Come on, let's get going!"

They soon realized that there was no road network here either, at least, not in the usual sense. What they used as streets was probably just the more or less random sequence of gaps and free spaces between the various facilities. It was often impossible to tell where the machine areas ended and the unused spaces began. Tower-like buildings stood in the open spaces, often very close together, forming a landscape that looked remarkably like a forest. Their progress was then more like a slalom than a purposeful advance. Sometimes they had to wend their way between stretched nets, sometimes they came across areas where the pear-shaped objects that had caught their attention earlier were densely packed in rows and files. They rarely found closed buildings.

Al made every effort not to lose his bearings. On several occasions, he only managed to do this by comparing the position of the sun with the time on his watch. A compass would have been useful here. This only occurred to him in passing, but it led him to a somewhat unorthodox train of thought. It struck him just how much a few tools would have helped them. But not just simple everyday objects. No, real purpose-made tools, with which he could, if necessary, intervene in a decisive way in the environment! Never before had he been so aware of the inadequacy of the means available to him. Never before had he felt so clearly that he was dependent on the environment, rather than the other way around. The fact that his path depended more on chance than on his own choices made this clear. He couldn't help thinking of the remote-control box, and the way the balls roll down an inclined plane, blindly bumping into obstacles, wandering past holes and gates, and finally ending up in the container at the bottom, no matter what has happened to them before.

Once again, they stood before an obstacle, a huge structure that extended so far to the right and left that it would have taken them an age to walk around it.

"Looks like a factory," Don noted.

Like most of the buildings, this one also willingly revealed its interior. Only in a few places was it enclosed by walls, and even these were made of that transparent material they casually referred to as glass. The roof was also transparent.

Intrigued, René approached an opening and followed a kind of path that seemed to lead through the interior.

"We might be able to walk through," he suggested. Secretly, he was keen to visit one of these facilities.

"Why not?" said Don, thereby giving the signal to enter the factory.

It really did seem to be a path, as the flat strip they were following ran continuously between the individual built-up parts. It was about a meter wide, without steps, often inclined, but never so steep that it would prevent them walking. The inclines were necessary to overcome considerable differences in height. The machines – if they were in fact machines – were of immense size and often several stories high.

"What might these be good for?" Don asked.

They were walking along behind a railing. Deep below, a whole series of sloping channels ran to gate-like openings.

"Perhaps a conveyor system?" René guessed. He couldn't hide his enthusiasm. "Magnificent. I would have loved to see this in operation!"

"Why do you say it's a conveyor system?" Don asked.

"Something is supposed to slide or drain down through the channels," René explained. "There seems to be some kind of sorting device up there, and something happens to the processed stuff when it gets down below." He was gesturing rather wildly, eager to make himself understood to the others, who clearly knew nothing about these technical matters. "Of course, all this runs fully automatically."

"Let's try it out!" said Al, and before anyone could stop him, he threw one of the pebbles that were still in his pocket into the nearest channel. No one would have stopped him anyway. Don didn't think much of excessive caution, René would have taken even greater risks to get the machinery going, and Katja hadn't even been listening.

The stone clanked metallically down below, bounced up once or twice, and then fell into a chute, where it rolled down and slid into the opening at the bottom of the track.

Suddenly, a note sounded in the air; a steady, high, bright, singing note, constant in pitch and volume. The four of them stood in wonder ... twelve large gleaming balls rose like soap bubbles from twelve funnels, spheres carried on a silky web of electric discharges that surged up and down like fountains in the wind. Pistons pounded and flywheels began to turn where the stone had disappeared. There was a dull grinding sound. The suite of actions ran through the machine parts like a wave, making wheels turn, joints bend, shafts rotate, relays click, and sparks crackle.

Katja suddenly awakened from her daydream and screamed: "There it is! There!"

The stone had reappeared. It bounced across a shaking conveyor belt. Then a fine-meshed wire structure like a hand reached over it, lifted it up, and tipped it into a pit where they could no longer see it from their standpoint. They rushed along the path and stopped suddenly: dry, breath-stopping air came up from below and hit them. Short-lasting, muffled hissing sounds became audible and the mild blue of Cerenkov radiation glowed somewhere at an indeterminate depth. Then everything went quiet.

"Nuclear decay," whispered René. "God damn it! Splitting the atom!"

* * *

They set off again in silence. Several times the path divided and they always chose the direction that would get them to the other side the most quickly.

Don, who was leading, suddenly raised his hand.

"Stop! Quiet!"

He pushed the others back.

"Too bad. We should have expected this. It's Jak with his followers. They mustn't see us."

They retreated behind a pyramid-like structure.

"They must have heard the noise," Al remarked. Don peered around the corner. "There they are, Jak and Heiko. And there's Tonio!"

"Are they getting closer?" asked Kat.

"They're deliberating," whispered Don. He turned around. "This is a good opportunity for us, but we mustn't be seen. We should watch, then follow them."

Don peeked around the corner again. "Careful, they're coming. We have to hide!"

René pointed upwards.

"Best up there."

The steps leading up were clearly not intended for climbing, but they managed by giving each other a hand. They reached a horizontal surface of about four by four meters, pierced by square holes. It looked vaguely like a sewer grate. Here, they could not be seen from below.

"Lie down," said Don softly.

They lay down on the hard surface.

"How uncomfortable," groaned Katja.

René looked into one of the holes, but saw only darkness.

"I hope it's not a gas outlet," he muttered.

"It's not in use anyway!" Don hissed.

Steps echoed from below. The second group must have been right beneath them. Now they could make out voices.

"... definitely from here. I heard it very clearly!"

"But what could it have been?"

"Maybe Don's lurking around here."

"Do you see any trace ..."

The voices faded. Only a few fragments of words were comprehensible.

"... search everything thoroughly ..."

Again, Don peered down. He held up his hand in warning.

"We can't do anything but stay here for now!"

"Shouldn't we ...", Al paused almost as soon as he had begun to speak.

"What?" asked Don, without enthusiasm.

"Communicate with the others?"

"What? What are you saying?"

"You heard me right: communicate with the others!"

"Have you lost your mind? Are you insane?" Don was beside himself.

Katja had propped her chin on her hands and was listening to him, half amused, half bored. Don went on talking for a while and Al let him finish.

"If we act together," Al tried to explain, "we can achieve more ..."

"And what exactly are we supposed to achieve? How are we supposed to be the first if we do that? Act together! You're talking nonsense!"

"Don, don't you understand? This is about much more than just being first. There's a mystery here that we can solve. Here we can clarify problems that affect the whole of humanity. Here we have ..."

"Be quiet, Al, please!" Don said firmly. Al glanced at the others. René sniffed suspiciously at the openings in the ground. Katja rolled casually onto her back, resting her head in her hands, and squinted at the light coming from above. It was no longer the deep blue of the day, but the inky colors of late afternoon, painting spots with rainbow edges up above.

"It's getting late," she said.

"Late?" Al repeated. "Maybe it is – too late."

"That's enough," Don warned. "Are you still with us or not? You know you don't need to stay here. Well?"

"Alright," said Al. He made a face as if he had bitten into a lemon.

"Okay then," said Don satisfied. He moved back to the edge and peered down. "They're over there. No doubt discussing what to do."

"The floor's too hard for me here. I want to go down," said Katja and got up. But Don moved swiftly across to her and pulled her back to the ground.

"Damn! Hold onto me ... !"

Katja hit the floor quite hard and stifled a groan.

"Al, help me!"

"Let her go, Don," said Al threateningly.

Don looked at him angrily. Al was no less angry.

"Let her go!" he demanded again.

"What's it to you?" Don snapped.

"Do something, Al!" Katja pleaded. She writhed under Don's iron grip. "Let's stop all this, Al! Let's give up! We can ..."

Don put his hand over her mouth. Lying down, Al grabbed his arm and pulled it around. Don let go of Kat and hit Al, once, twice, without getting up. But Al caught his fist and turned around. Then suddenly they flew apart. René had thrown himself between them.

"Stop. Be quiet. Just be quiet!" They listened intently. They could hear footsteps, then voices.

"... somewhere over here ..."

"... sure I wasn't mistaken ..."

The footsteps clattered away. The voices became unintelligible.

"Stop your damned quarreling ...," scolded René. "You're going to mess everything up! If you're going to do that, why are we even trying?"

Don and Al had calmed down a bit, but occasionally threw angry glances at each other. They waited for a while and listened. Several times the footsteps got louder, before growing quieter again.

Evening came and then night. It fell quickly, as always when the sky is clear. The reflections faded. The machines lost their sheen. Their edges softened and their corners dulled.

"Right, let's climb down!" Don ordered. "We have to follow on their heels, otherwise they'll slip away from us!"

They helped each other down and soon arrived safely at the bottom. Sounds came from afar.

"Over there!" whispered René. "Come on, quickly!"

They tiptoed along the path, which wound like a light gray ribbon between the sleeping mechanical monsters. Katja was making little effort and suddenly stumbled, only catching herself at the last moment on a kind of vertical, spiral-shaped metal structure. The shock set its elastic material vibrating and was reflected back and forth. There was a sound like a snapping string which faded, then repeated with each passing echo, gradually growing quieter, but still disturbing the silence of the hall.

There were voices, noises, impacts of shoes on metal ... Don looked around wildly.

"Over here!"

He jumped off the path onto a wide ledge that jutted out from a humped metal construction and crawled over the curved surface, a shadow in the darkness ...

The footsteps were approaching quickly.

Al jumped down to the step – it was only about a meter below – and held out his hand to Katja and René. Following Don, they worked their way around the metal ledge.

There were three shadows somewhere above them.

Don was now far ahead of them and they had to rush on just to keep him in sight. He swung himself over a barrier. Then it happened – a kind of rake swept along the track and Don let out a shrill scream.

Suddenly the air began to sound out, singing piercingly, without interruption, without wavering.

A bright blue light raced across all the surfaces, along all the edges, catching on the corners and tips, hanging in the rows of wires, while twelve blue-white balls of lightning pulsed in the same, metronome-like rhythm.

An awakening, a stirring, a movement spread as if carried by a breeze. There was a pounding of pistons, a whirring of flywheels, a groaning of chains, a crackling firework display of sparks that rushed up to drown out Don's screams.

Three figures stood motionless at the top of the path, bleached by the blue brightness, in front of a dazzling curtain of lights and receding shadows.

"Nuclear decay," groaned René.

He ran alongside the conveyor belt that Don had stepped on, saw how he disappeared into a dark hole, pushed himself up onto the path, stormed on, then saw how Don reappeared – a shovel pushed him over a series of sieves, until he rolled through the wide meshes of the last into one of the chutes that they had already seen from the hatch. Tumbling, desperately trying to take hold, he slid down the slope, dived into an opening, and was spat out further on, his body jerking over the shaking mechanism until he came to a bell-like structure where a kind of wire net was cast over him, throwing him up and tipping him into the pit where they had seen the blue glow a few hours before.

René had stopped by now. He pressed his hands against his face.

The whole process stopped as quickly as it had started. Everything went quiet. Deadly quiet.

After the abundance of light, the twilight took a while to breathe its soothing grayness back into the world.

Footsteps could be heard on the path above. Jak, Tonio, and Heiko moved off, without caring about the other group.

René looked for Al and Katja, and found each standing alone and lost beside the conveyor belt. They too left the room silently and started to make their way back to their campsite behind the wall under the sparkle of foreign constellations.

* * *

The three of them paid no attention to the stars. They were worn out and stressed, and even if they did begrudge Don his mistake, they still felt strangely insecure without the authority of their companion, who had always known how to act immediately and swept the others along. Above all, they were in a state of awe when they thought of the tremendous forces that lay hidden in all those batteries, capacitors, wires, tubes, and containers – forces that only needed a trivial trigger to set about performing their task, mechanically, regardless of whether it still made any sense.

Although the night was as bright as usual, it was not easy to find their way back. It took them half an hour before they arrived at the wall. Ten minutes later they were lying on their rubber mattresses in the tent. They pulled their sleeping bags over their ears so that they could see nothing and hear nothing. This was the lesson from a million years of the development of self-awareness, turning away from nature and adapting to an artificial environment, and yet still animal-like in their endeavor to hide.

"Al!"

Katja whispered so as not to wake René, who was tossing and turning restlessly in his sleep.

"What's up, Kat?"

"I've asked you something twice already."

Al sighed.

"I know."

"Do I mean so little to you?"

"But Katja, try to understand. This is …"

"Al, we don't need to struggle anymore. We don't need to get upset and torment ourselves …"

Al tried to interrupt her: "Please, Kat, listen to me …"

Katja went on undeterred.

"Maybe it's not too late to re-register. Maybe we'll be admitted as a couple, and then … Don't you want that?"

"Yes, Katja, but …"

"So let's leave this silly planet and all these boring machines, this whole lousy city …"

She had raised her voice and Al hissed "Shh," as René rolled over, grunting.

"Al, I've only stayed this long because of you. You don't understand how terrible everything is for me here. Just think how wonderful it could be! We would play with the remote-control box together, let ourselves drift through the three-dimensional rooms, experience the color melodies and the stereos! Why don't we just give all this up together?"

"We must wait until this is over, Kat! If only you could understand!"

"Won't you give it up for me? I can't wait until later. I don't want 'afterwards.' Right now!"

Al was silent.

"Won't you give up?" Katja insisted.

"No," said Al, "but ..."

"You don't need to say anything more. That's enough for me," said Katja harshly. She retreated, rolled herself into her sleeping bag in the farthest corner of the tent, and went absolutely silent.

The next morning, they were rudely awakened. Someone had lifted the tent flap, letting in the dazzling morning light, and a voice called out: "Hey, you sleepyheads! Out of bed! Out with you!" Katja slipped out of the covers, hopped between the barely roused bodies of Al and René, and threw herself into Don's arms.

"Hello, Kat! So, what do you say about that? Al, René, you sleepyheads!"

Al sat up sleepily.

"Where did you come from?"

"Out with you, come on!" Don cried in great spirits. "You see, nothing happened to me!"

Al struggled out of his sleeping bag and crawled awkwardly out of the tent. He punched Don playfully in the side. For the moment, he forgot about their quarrels.

"Well, tell us what happened."

By then, René had joined them.

"So," Don began, "just as I was about to walk across the conveyor surface, a kind of huge rake came flying up and carried me into a passageway. After that, it wasn't much different from the test yesterday morning – there were some stops in which I was irradiated, sprayed, blown on and whatever – then I was swept downhill on a slippery surface. I was just a plaything, rolled over and over, shoveled into a kind of funnel, then a journey over a bumpy surface. I can tell you, I had my stomach in my mouth, I was shaken up so much. Then, in the end, something pushed me up high and let me fall back down again. I landed softly in a kind of stretchy tissue. I just slid gently down in a spiral and the net gave way. So, there I was, outside. That's all!"

"But what did you ..."

René stopped himself asking the question, but everyone knew what he wanted to ask, just as they knew that Don was trying to disguise it with all this breezy affability. The question was: What had he been doing all night? It was a serious breach of rules to sneak away during a group mission, especially for a whole night. It was surely no coincidence that Don was now so well rested and in good spirits. But they remembered that they too had not always behaved entirely according to the rules. And there was an excuse in the present case: his experience had undoubtedly been something completely out of the ordinary. Don's screams of terror were still ringing in their ears. So, they remained silent.

"Well? Why are you just standing there?" asked Don accusingly. "Isn't it incredible?! How do you explain that?"

"Some things are perhaps not so mysterious," said René. "It seems to be a facility for atom smashing, a kind of matter converter. Put simply, it works according to the following scheme: you put matter in at the front and it comes out at the back, but in another form, the one you wished for."

"Very convenient," said Don.

"The actual conversion only takes place in the atomic reactor – that's the part of the machine where the blue light comes out. It's Cerenkov radiation. It occurs when electrons or other charged particles shoot through substances very quickly – during nuclear decay, for example. Everything that happens before is only for analyzing and sorting the material …"

"Just what I said," Don interjected.

"The results of the analysis are then used to properly dose the effects – alpha particles, slow neutrons, gamma rays, and so on. The nuclear reactions leading to the desired result take place in the final reactor."

"So, why wasn't I turned into gold long ago?" Don asked.

"Some kind of safety device is probably built in," René speculated.

"That's almost certainly it," Al confirmed. "On Earth, this sort of super machine is always designed so that it can't hurt anyone."

"Great," said Don, "that means we don't need to be particularly wary of them and we can focus on Jak. I have something special in mind. Please, pay attention!"

With short, hurried words, he explained his plan to his companions.

Sometime later, they were walking along the city wall, following the same route as a few days earlier. They felt strangely ambivalent seeing the castle buildings on the left, so colorful and three-dimensional, as if they could reach out and touch them, and at the same time knowing that it was all just an illusion. But even apart from that, this striking clash between the Middle Ages and all the paraphernalia of an apparently utopian technoid civilization was strange enough, and it did much to reinforce this impression that the whole situation was unreal.

They headed to the square from which the gate led to the drawbridge and got there unchallenged. Walking under its wide arch, they turned to the door that led from the back into the building. René had already been there and led the way. They went up a flight of stairs, past several doors, climbing hurriedly up the spiral of worn stone steps. Finally, they came to an old carved door. It was open, and René stopped. They had reached their destination. The room was an armory. In the middle there stood suits of armor, on the walls were instruments of torture and weapons of various kinds, some familiar, others not.

Al returned to the stairs and climbed a floor higher. He was able to squeeze through a hatch onto the flat roof, with dust and sand flowing down on him in rivulets. It was surrounded on all sides by a chest-high defense, with cuboid blocks built on it at regular intervals, apparently to protect against projectiles falling from the side. Some canisters of grapeshot stood on wheels by the parapet. The many fresh streaks in the dust indicated that they had been pushed there recently. Indeed, Jak had used them to fire shots. From up here, taking aim was a straightforward matter. One could see far into the countryside in all directions, over the rust-red and jade-green pattern of rooftops and gutters, over the twisting and bending line of the city wall, over the moat, over the bridge – the bridge that ran without interruption across the water and was spanned by a high arch on the other side.

Al went back down to the armory. Markings on the wall revealed that weapons had been hanging there for a long time before Jak and his people had taken them down. Don was already weighing up a saber in his hands.

"Pick something suitable!" he called out.

"Make sure you don't choose anything broken," René warned. Don was rummaging in the drawers for ammunition. Triumphantly, he pulled out a box of bullets and a bag of powder.

"René, can you tell me how this works?"

René looked attentively at the items Don was showing him.

"They seem to be weapons from different centuries," he said. "I think this is the most modern." He pointed to a device that resembled a pistol, but was considerably larger. "Here's the ammunition for it." The others gathered round him so as not to miss anything. He inserted a thumb-sized cylindrical shell into an opening, then closed it with a flap and went over to the window.

"Watch out! I'm going to give it a try!"

The weapon had a handle where the trigger is usually located on common pistols. René stuck it out of the window and pulled the trigger with his index finger. There was a loud bang, and then a crashing sound. René stood shrouded in a cloud of smoke, but before the smoke obscured their view, they had all observed what had happened: a circular hole had been made in the courtyard below. A few cobblestones had shot off in different directions and a white cloud of smoke was left to drift away.

"Exploding ammunition," René said approvingly.

"Excellent," said Don. "Each of you take one of these. And enough ammunition. Maybe we can use some of the other stuff too."

They spent a while longer trying things out and eventually they had all equipped themselves with what they liked best. Don was rather taken with the handy firearms and their exploding ammunition, which they called pistols for short. He wore a belt with a pistol and carried a weapon like a mace. Al had one of the pistols and René two. Katja found these weapons too heavy. In the end, she attached a delicate, gold-studded dagger to a loop on her jacket.

"Now we're ready," Don announced. "We can pay Jak back for what he did to us!"

* * *

They hadn't brought a ladder this time so they had to go back to the platform. This was a minor setback, but it did have its advantages: they already knew the way from there, and besides, it meant they kept open the option of retreating directly to their temporary camp at any time. Having had plenty of practice by now, they cheerfully climbed down the ladder without paying any attention to the strange optical illusions.

Carrying their ancient weapons, they presented a peculiar sight among the ultra-modern buildings and machines. It's actually crazy, Al thought; have we become so lazy that we can't take anything seriously anymore? That there's nothing left for us but pleasure and entertainment? That we voluntarily give up anything that could take us a step further, because it would mean a small loss of pleasure and entertainment?

They silently avoided entering the matter transformation factory. They walked around it, although that lost them a bit of time, then veered to the right and entered a street that only Don had seen, and even he, only in the starlight.

"I must have come out somewhere around here," he said, pointing to a series of openings beneath which there lay semicircular hollows, as though ready to catch a falling object. They came up close to these and René bent down to pick something up. He couldn't hide his surprise. In his hand was a billiard ball-sized object with a smooth surface, but not perfectly round, two sides being slightly flattened.

"The pebble!" Al exclaimed. "Yes, my pebble. The one I threw into one of those channels yesterday!"

René turned it over in his hand, squeezed it gently, and smelled it.

"Right, Al, it is the pebble. But it's made of sulfur! Unbelievable!"

The otherwise inconspicuous object was passed from hand to hand. "They certainly knew what they were doing," Don praised. "But let's be getting on! There's no time to linger. It's already past noon."

The further they went, the more densely the buildings stood. The pillars, masts, and towers seemed to be moving together and there was less space to move around on the ground. The building now strongly resembled a gigantic substation from the time when electricity was still the main means of energy distribution. Although the various metal, glass, and plastic structures with their rows of struts, wire grids, and frames, were certainly intended for other purposes than transformers, insulators, and cables, and although Don, Al, Kat, and René were now sure that the machines had the skill and predisposition to spare humans, they still moved a little hesitantly between them, as if there might at any moment be some kind of destructive discharge. And so, they continued somewhat awkwardly on their way.

The hill was not far off.

"It really is a hill," said Don. "The paths do actually go uphill."

"That doesn't mean it has to be a hill," Al contradicted. "It could just as well be a gigantic building. The paths could lead up to the roof – but they're probably not even paths."

"What else could it be?" Don growled.

"Free space to build, improve, and repair."

René shook his head thoughtfully.

"If it is a building, then it has a special status. It's shielded from the outside and the roof is opaque."

They had reached the point where the strips of land they were using as paths began to rise. Since there was no other way to continue, they began to climb the gentle slope. There were still machines, devices, automatons, or whatever they might be on the right and left, but it was now clear that other kinds of structure prevailed. These were less compact – scaffolding made of thin rods, widely stretched nets, and towering pillars, all connected high above by a pattern of gray threads that merged with the sky.

"Those look suspiciously like antennas," René muttered. Don picked up the remark.

"That would be consistent with our assumption that this is some kind of central hub."

"Then it can only be below us," said Al.

"There's a door," cried Katja, who had so far been walking silently next to Don.

They hesitated instinctively.

"Do you think Jak is inside?" Al turned to Don.

"Seems likely," Don replied.

"Is it advisable to just march in?" René asked. "It might be a trap?"

Don unbuckled the pistol from his belt.

"We are armed. Get your guns ready!"

Now, without hesitation, he stepped through the circular opening in the sloping wall which Kat had referred to as a door. The corridor they entered also had a circular cross-section, with a diameter of about three meters. There was a strip of a glowing material embedded in its ceiling, from which a soft light emanated. After only a few meters, they came to an intersection in the form of a small room.

"They have eyes here too," grumbled René, pointing to dark hemispherical glass lenses protruding from the walls.

"There are control rooms here," Al called out softly. He had taken a look down the corridor on the right. It expanded quickly to become a large hall like a gigantic ellipsoidal bubble. The floor or, to be more precise, the lowest, approximately horizontal parts of its boundary surface were occupied by block-like rows of control desks, each embedded with thousands of buttons, levers, dials, and the like.

"The main control center," René whispered in awe, with something that came close to reverence in his voice. "The heart of the city."

Don was soon standing between the control desks, directing his attention forward in order to detect Jak and his two companions as quickly as possible should they appear. As he walked, René was staring at the hieroglyphic-like characters engraved next to the switches, and almost slipped over on the unfamiliar curved floor. Meanwhile, Al divided his attention between the location and the facility and Kat tried to distract herself from the uncomfortable situation by imagining fragrant combinations of scents.

In this way, they covered a considerable distance without seeing any sign of Jak. They crossed halls, most of which were empty except for switching devices, while others contained screen-like objects, wire frames, and the like, mostly attached to the walls or embedded in them. They walked along endless corridors and entered dome-like rooms, from which they could reach higher floors via spiraling ramps.

René finally found a clue: in one hall, a wall had been lifted to reveal electrical circuits. They contained many unknown elements, but they were still recognizable as electrical circuits.

"They must be nearby," whispered Don. Tensely, he crept through the hall into the adjoining section of corridor, then through to the entrance of the next hall. There they stood, the tall Jak in his beige combination with white boots and white cap, Tonio, medium height, slim, and black-haired, dressed all in blue, and Heiko with his blonde brush haircut in grey flared trousers and a short black jacket. They were moving about quite unconcernedly and talking loudly, but the echoes were so prominent that nothing could be made out. They were not focusing on the switchboards, but working on the wall.

"We'll soon spoil that for them," whispered Don. "Al, René, when I count to three, we all shoot together. Then run through the hall to the other exit immediately, so fast that they don't even come to their senses – unless we've already hit them. Katja, did you hear?"

"Yes, of course!"

Quietly, they moved a little further forward to get a clear line of fire.

"So, get ready: one, two, three!"

The shots rang out, clouds of smoke rose, and splinters whirred past them. Al and René charged forward.

When the smoke had cleared, they saw a body on the ground, torn into several pieces, its parts crushed and twisted.

"The others are behind the control desks," Don called to Al and René. "Reload immediately so that we always have one pistol operational. We'll just aim at the wall."

He fired again and a shot also came from the other side.

Don seemed to be beside himself with joy.

"Who would have thought that, Jak, old boy!" he cried. "What do you say now?!"

He waited a few seconds, then turned to Kat: "They don't want to give away their positions." Then he called to the others: "Watch out, Al and I are going down. Kat and René cover us. Understood?"

Then something hummed behind him, and a light breeze came out of the tunnel. A round surface appeared, flashing dimly at the top, with openings, screens, and the like below. A mechanical arm reached out and two soft but firm pincers closed around his middle. He struggled for a moment, before finding himself deposited gently on soft cushions. The next moment Katja was sitting next to him. A further short glide and a jolt and René floated in. Another barely noticeable jolt, and Al was with them too.

They were inside a kind of hovercraft. They glimpsed Jak and Heiko rising from behind a table, then they could already see the blurred grey of the corridor above them. The strip of light seemed to wriggle slightly, then daylight, the sun, blue sky, objects whizzing by, metal, plastic, glass …

They were carried swiftly along, unable to do anything about it, without stopping, without hesitation, to the city wall, to the very spot where their ladder hung. Here the sliding door opened. They got out. The door slid shut and the craft set in motion and darted off, like a flashing reflex in the city's forest of machines.

They had no idea what was going on. They just stood there and watched it go.

* * *

A moment later, Don hurled his firearm at the wall and began to curse softly but unrestrainedly regarding his choice of language. When he finally noticed that the others were not participating in his outburst, he cursed them too, but without achieving any visible effect. René stood on the bottom rung of the ladder and swayed from side to side with an absent expression on his face. Katja held her dagger in her hand and appeared to be cleaning her fingernails with it. Al watched Don and grinned.

"One would think there's nothing funnier than one defeat after another," Don snapped at him.

"We were so close to our goal and again nothing! These damned machines! Why are they interfering? They just seem to help Jak! How did he make that happen?"

"I don't think they're taking sides," said Al. "That's just not the way machines operate."

"But the three of them were tinkering with the circuits! Maybe Jak managed to reprogram the machines differently, to get them on his side."

"Jak and Heiko were just as surprised as we were," said René from the bottom of the ladder.

"So why did they attack us?" Al raised his hand in denial.

"They didn't harm us. Their intervention was not against us personally, but against the attacker. The moment they realized we were planning something destructive, they prevented us."

"But they let Jak and his people tinker with the circuits. How do you explain that? Remember how quickly they were there when René covered the eye?"

"How could I possibly explain that? Perhaps no precautions were made against changes in the control center – otherwise the city dwellers would not have been able to set up anything new."

Katja threw the dagger aside.

"I'm going back to the tent," she said, "will you come soon, Don?" Then she turned to René: "Can I get past, please!"

René jumped off the ladder, and Katja climbed up slowly. She could feel exactly how the fabric of her pants pressed tightly against her skin with every movement, and she enjoyed imagining the vivid impression she was making below. René let out a soft whistle. The three men were left gawking until Katja disappeared behind the parapet.

Don took a few aimless steps.

"I've had enough for today," he said. He cleared his throat, then quickly climbed up the ladder.

"Horrible," said René, turning on his heel and looking over the city.

"Sometimes I'm not sure if this is real," he said.

Al was trying not to think about Katja and was happy to respond to his companion's remark. He immediately knew what René meant. The city area lay before them like some kind of incomprehensible abstract picture, with a pattern that just did not reveal its meaning; silent and static, a symphony in lead gray, ivory, and silver.

"I wonder if the illusion doesn't go much further than we think," said René, and suddenly he added with strange urgency, "Al, are you sure that there is anything here at all? I mean, around us?"

"Well, yes," Al reassured him. "What you feel must exist, and so must what you see and hear. And everything else. Maybe it's a little different than you imagine, but something is undoubtedly there. And the beauty of it is that it not only exists, but it affects its surroundings, and the surroundings affect it back. It influences the future and is itself a consequence of the past. It contains potentials. Possibilities are awake in it; energies are waiting to be triggered; and often, perhaps more often than we can perceive, something in it comes alive – in some form, although not necessarily in the same way that we are alive." Al fell silent for a while and then went on: "And you see, René, that's probably why this planet fascinates me much more than all the others I've seen so far. This is an opportunity – a truly wonderful opportunity – to learn a little more about what exists in the world besides us. Of course, we can never check directly that things really are as we experience them – we always need waves and vibrations and impulses to see, to hear, to feel – we can't break free of that. But we are free to search in different directions. We will never understand the absolute – only the connections. For us there is nothing absolute, and perhaps the absolute is just a pipe dream, a fiction. But for us there are connections. They are our reality."

René didn't quite understand what his companion was saying, but he felt that maybe he didn't really need to understand it, and that he could be satisfied with things being as Al described them.

"So, what should we do next?" he asked. "What you said about the machines makes sense to me. But will they even allow us to continue our research?"

"As long as we don't use force, certainly."

René shrugged his shoulders. Although the evening was drawing in, the weather was as beautiful as ever, the air fragrant, and the temperature mild. There was nothing particularly natural about this weather anymore, René thought. He turned back to Al.

"Don't you find it rather eerie to face these nebulous and unpredictable forces?"

"They're not that nebulous," said Al. "I would say they're pretty easy to see through if you just have the key. You can understand them – not their technology, but their behavior. What I mean is ..." He fell silent.

"You mean to say they obey simple rules, like the four classic laws of robotics?"

He began to recite: "First, the robot has to protect humans and prevent them from being harmed.

"Second, it has to obey humans.

"Third, it has to make sure that it itself is not damaged.

"And fourth, it must always behave in such a way as to destroy as little as possible of its surroundings."

"Actually, I was thinking of something else," Al replied. "I wanted to say that ... I find it hard to express. I think that's not the whole thing. There's something else behind it that we haven't discovered yet." He was clearly finding it hard to drop these musings, unproductive though they were. "But in any case, the laws are surely correct. I am convinced that all the machines that have been built somewhere or are still being built will obey such rules. Anyone who gets far enough to be able to build them will also be sensible enough to protect themself from them. But that raises a whole lot of questions. What happens if the beings that built the robots have gone extinct? Can these then change their programs on their own? And what would be the basic rules for robots then? The beings of this planet may also have had a different set of ethical values to us. To give just one example, they might have inserted a rule to protect all living things, between the first and second of our laws."

"I don't believe that," René countered, "because it looks like they wiped out all the animals."

"That may be so," said Al, without commenting further. "After all, they may have had a more complicated legal system than ours. But I don't think that's the important thing, because in principle there could have been no other goal than to make machines that would protect their builders, that would owe them obedience; and that they would not be allowed to damage themselves or each other. But now comes what is unclear to me: How different are we from the intelligent beings of this planet? Or put another way: Do the robots see us as their masters? And there is another obvious possibility: Do they consider us as robots – as colleagues, so to speak?"

René snapped his fingers in surprise.

"Indeed, that is the most likely thing!"

"Exactly," said Al. "They've examined us closely. But what means do they have to distinguish an intelligent creature from a robot? That we must be one of the two was surely clear to them from our rational reactions in the tests. But what conclusion would they have reached? We don't understand the local language. We can't give them instructions. And above all, we must look quite different to those that built them!"

"We don't know their technology and they don't know ours. That's the problem," said René. "We're robots to them, and they treat us as such. We should be glad they didn't destroy us! But how should we behave then?"

"As long as we proceed without violence, I think there's nothing to fear. But I also think we won't get very far. Their main duty is to protect the inhabitants of this planet, and they'll try to do that even if those inhabitants are long dead. It'll undoubtedly be a priority over the rule of preserving other robots. As soon as we get close to the secret, we should expect no further kindness from them."

"Al," asked René, "are you still bothered about trying to beat Jak?"

Al turned to him and looked him in the eye. Now he understands me, he thought, now he finally understands me.

"Jak is completely indifferent to me, and I frankly don't care whether we win or lose. Even the physical nature of the inhabitants is of little interest to me. I want to know something completely different – more urgently than anything." He lowered his voice, as if he wanted to tell René a secret. "I want to know what happened to them. Because it's almost certainly what will also happen to us one day."

They were silent for a few minutes and watched as the fuzzy evening glow suffused the glass and plastic surfaces with a mysterious, almost promising shimmer. Behind all that lay transparent and clear before them, there lurked something mysterious. They looked at each other, and it dawned on them that their task had just reached epic proportions.

"How should we go on from here?" René asked again. "Is there even a way forward, given how helpless we are right now? Can you think of a way?"

"There is a chance if we manage to set aside all these childish and silly things – sporting rules and agreements that may be good for other places and other purposes, but don't seem to be working for us here. And then we must use all the means at our disposal. It'll be hard, because we haven't had to do anything like that for millennia. We assumed there would be no more tasks like that, or they just didn't interest us. But," he added thoughtfully, "it'll be difficult, and it may also take a long time."

He fell silent. With a touch of emotion, he saw how expectantly René was looking at him. He said, "There could be another way!"

"What's that?" asked René.

"We could try to communicate with the machines," said Al.

"Then we'd find out everything we need to know, without effort," murmured René with new confidence.

"Maybe," Al qualified, but this 'maybe' expressed the greatest hope of his life.

* * *

After another night in the tent, a night full of expectant dreams, they were woken up by Don shifting noisily and crawling out of the tent.

"Do you think you can convince him of our plan?" asked René quietly.

Al stretched and yawned.

"It'll be difficult, but I'll try."

"What are you whispering about?" murmured Katja, half asleep. "What time is it?"

"Time to get up," said Al and left the tent behind René.

A little later they had gathered below, at the foot of the ladder.

"Do you still have your weapons?" asked Don.

"Are you going to try another shootout?" asked René.

"Of course! Do you think I'm going to give up so quickly? After all, we've already eliminated one of them. There are only two left. Of course, we have to be careful. We must smash all lenses in the area and use the time during which the machines can't track what we're doing. Does anyone object?"

René nudged Al secretly. "The plan is good," he said, "but we reconsidered it yesterday. We have another plan. Tell him, Al!"

Al did so, and Don listened for a while, frowning. Then he interrupted the explanations by ostentatiously covering his ears.

"What strange ideas you come up with when left alone!" he exclaimed. "Are we going to start studying now? Linguistics and philosophy and that kind of stuff? Just remember that a goal is best achieved by going straight for it! I want something – and I take it. That's how to achieve a goal."

"But what exactly have you achieved so far?" asked René.

Don now adopted a didactic, persuasive tone.

"Look, guys, I'm not claiming that you wouldn't achieve anything your way. But it would take far too long. In the meantime, Jak and Heiko will have reached the goal, and we can kiss goodbye to victory."

"Now you listen, Don," insisted Al worriedly, "you must realize that we have to use other means here. Look at this thing!" He pulled the pistol from Don's belt and held it under his nose. "It's with toys like this that you want to defend yourself against machines, against an intelligence that you don't understand at all, an intelligence that has produced all this?" He dropped the weapon onto the plastic floor in front of Don's feet, grabbed him by the shoulder and turned him round forcibly. "Look at this huge factory! But I want to tell you something else: this is just nothing. It's much too simple. It's different from our own systems – it's built differently and set up differently – but it's still frighteningly similar to them. This is at the same level of development as our own technology. But with these guys, you know, they've come way past that! They're much further ahead. So, there is something else somewhere, something that came later – something that belongs to a higher level! It must be somewhere! And it's much more complicated and powerful than you can imagine. It makes your intentions and your plans simply laughable!"

Don had never seen Al so excited. He was a little confused. He heard him ranting, and he didn't understand him.

"But Jak," he stammered. "Jak and Heiko …"

"… are failing just like us! They're messing around with the circuits of the central control room – it's just nonsense! In fact, they may well ruin our last chance."

He suddenly fell silent. In his eagerness, he had not been paying attention to what was going on around him, but now he saw something in Don's face … reflected in it was surprise, dismay, horror. Don's eyes were fixed on a distant point, on something that Al couldn't see, but he knew, even without seeing it, that it was something breathtaking. He turned around to look at the wall.

There was no wall. There was no ring of old buildings and no belt of ivory-colored villas. There were no meadows, no rocks, hills, or lakes, and there was no wall of mountains, no horizon. They were standing on a disk, and this disk ended in front of them. The sky reached below the horizon line, which had disappeared, blue, radiant, without the smallest hint of cloud, without the slightest haze.

For a second, they were petrified, then recoiled.

"It can only be an optical illusion," Al screamed, but he drew himself back with the others, so strong was the impression of the enormous abyss filled with sky.

"We have experienced something like this before," Don recalled. "Think of the bridge!"

"Go ahead," René shouted, "let's try it out. Let's see whether everything has disappeared all around or whether it actually continues!"

They yelled at each other, trying to kill the fear inside them, but it didn't work. Don took a step forward, toward the blue nothingness, but he was seized by such a violent choking dizziness that he convulsed, leaning face first against the cool, comfortingly solid wall, his arms raised as though he were being crucified.

"My God," said Al, "we can't let ourselves be defeated by a few manipulated light vibrations!"

"Maybe it's true," Katja sobbed. She ran to Al and buried her face on his shoulder, in the soft hollow between his collarbone and his neck.

"We have to give it a try," René shouted. "Give me a hard object!" Since no one moved, he ripped the dagger from the loop on Katja's jacket. "Al, hold my feet. I'm going to crawl forward. Come on, help me!"

Al pushed Katja aside and stepped over to René. Meanwhile, Don turned around slowly, as if sleepwalking.

René went up to about ten meters from the edge and lay down on his stomach. Al sat down behind him and held his companion's legs just above the ankles. They crawled forward, a few centimeters at a time. René stretched out his hand and tapped the ground with the handle of the dagger. At a snail's pace, they approached the step ... again and again the pommel of the dagger struck the rock-hard mass of the ground ...

Katja suddenly screamed. They stopped and looked around. Then they heard the wail of a siren and saw threatening columns of thick black smoke rising from the interior of the compound, floating high up, until they ended as if cut off. Something fiery, like a large glowing ball, now rose up somewhere behind the rooftops and began to move faster and faster until finally it disappeared into the air. By then, a new column of smoke had come into sight. It looked as if the disk which carried the factories was hanging freely under the sky by means of these columns. "Continue," René hissed through his teeth and crawled forward again. Al stayed close behind him.

René paused again.

"Do you see that?" he asked.

Al raised his head and peered over his companion's body.

"The edge is moving," he said.

Indeed, the edge was not lying still, but oscillating. Twenty centimeters were added, twenty centimeters fell away, in a constant wave motion.

"That's confirmation that it's just a trick of the light," Al shouted.

René pushed himself forward.

"Come on!" he shouted, impatiently tugging at Al's clinging hands with his legs.

The pulsations of the edge seemed to become restless. They moved further forward and further back, until suddenly the edge rushed closer and slid beneath them.

The disk was now behind them. It was paper thin. In front of them, next to them, and below them was the sky. They were floating in the sky. No, they were not floating, they were lying. They were lying on solid ground, where René's mechanical blows were still ringing. By now, Al had let go of René. His hands too touched and felt this ground, which could not be seen. And even though they knew it was just an optical illusion, the contradiction between what their eyes were telling them and what their sense of touch was telling them was simply terrifying. They had to close their eyes to keep from going crazy.

Behind them they heard Don and Katja calling out their names.

With their eyes still shut, they crawled towards these calls. Ever faster and more urgently, they rushed forward. They could have walked, but then they would have lost the close contact with the only thing that still allowed them to keep their wits together – the solid ground.

Suddenly, the screams sounded different. Hands reached out, they were shaken, and yet still they did not dare open their eyes. "It's over, do you hear. It's over!"

Al felt a soft, wet face against his, and only now did he raise his eyelids, although he was ready to close them again immediately. Katja knelt in front of him and kissed him. She was crying. The abyss had disappeared. The wall was there again, as were the wrinkled buildings, the darkened and patinated roofs and pinnacles, the gables, the bay windows, and the arches.

Nothing remained of the spectacle they had just witnessed.

Or rather, there was something! The black smoke over the machine area. It was no longer disposed in the form of columns. It had evolved into torn, twisted, fluttering structures that slowly crept southwards.

* * *

Katja was still kneeling on the ground in front of Al. René tried to get up. He was trembling with excitement. The handkerchief fell from his hands as he tried to wipe the sweat from his forehead. Katja was also in a state of shock – her face was drawn.

Don leapt towards them and pushed her aside.

"That's enough of your smooching!" He pushed Katja roughly aside, planted himself in front of Al, and raised his right hand, clenching it into a fist.

"I'll teach you to grope Kat with your dirty fingers."

He drew back in order to lash out at Al, who was crouching in front of him on the ground, but René jumped up and wrapped his arms around him from behind, firmly gripping Don's wrist. He turned around angrily. He was excited, but his face bore no traces of any particular experience. However, his behavior was indeed thought-provoking. It did not fit with what they had just been through.

René, who was looking Don full in the face, suddenly found the explanation.

"You've reduced your intensity of experience! You scoundrel." René was so indignant he was unable to go on speaking and had to take a deep breath to compose himself. "Shame on you, you coward! You have done it, haven't you? Just admit it! You're not really taking part anymore! – And how do you manage that? You're just sitting back and watching comfortably!"

Don forgot everything else in the face of this accusation. He turned pale and tried to justify himself, but his stammering was not convincing, as he could clearly tell by the looks the others were giving him. The usually quiet René was still under the shock of his recent adventure, and now he was going wild, letting out everything he had bottled up inside him and taking it out on Don, who was so overwhelmed by this unexpected onslaught that he sought in vain to reassert his usually effective powers of persuasion. He would not admit it, but for the others it was clear: he had failed in a difficult situation.

But the little group was not to see any respite, because something else was stirring among the machines. One of the hovercrafts appeared, disappeared behind a building, then re-emerged. It hung a little askew in the air and there was a clinking sound as if a rain of shards was falling on it – and that's exactly what was happening. The stern sliced a glass roof from one side of a hall to the other. There was a crash. Two masts were bent and sank to the side almost in slow motion. They then struck across the rooftops like whiplashes, leaving behind deep, jagged scars. Only two hundred meters from them, the hovercraft raced towards the wall and shattered into a white–gray fountain of disintegrating debris. Immediately afterwards, the shockwave from the impact hit their eardrums with painful force.

"The whole city is out of control," said René. "What on earth could have happened?"

So far, none of them had had time to worry about the causes of the uproar. But now that the question was asked, Al's brain immediately connected the current events with the recollection of their three opponents tinkering with the circuits.

"Come on!" he called. "Maybe we can still save something. It can only be due to problems in the central control room."

Everyone obeyed Al's call – René, because the fate of the city was now much more important to him than before; Don, because he was hoping for an opportunity to regain his leadership role; and Katja, because she didn't want to be left alone.

The city now presented a completely different picture to the one they knew from the previous days. René had wished he could see the machines in operation. Now this wish had come true – so fully that it went well beyond even his technically oriented inclinations. Literally everywhere, machines whizzed, hummed, howled, and crashed, steam hissed from nozzles, liquids bubbled into containers, hot gases rose shimmering in the sunshine, and the air smelled suffocatingly of sulfur dioxide and ozone. Wheels turned, centrifuges rotated, belts ran, chains rattled, cranes swiveled back and forth, gates swung open and shut, carts rolled over rails, stopping to let sand trickle in, then starting again, stopping to tip it out, rolling on and on, round in circles, ready for further trips on crisscrossing stretches of track. Gripping tongs grabbed things, passed them on, laid them on tables, clamped them, and set them in rapid rotation, while drills stabbed, band saws cut, rods and bars were twisted into spirals, blows fell with a dull thump on sheet metal strips running underneath, hammers pounded, tongs grabbed once more, passed things on, pushed them into place, grasped them, moved them on, shoved them, and so it went.

Only the smallest part of what was happening resembled any known processes in chemical factories, technical manufacturing facilities, or electric power plants, and even this part apparently served no meaningful purpose. Plates were first punched, bent, welded together, filed down, and sprayed, then at the following stations disassembled, cut up, annealed, and reduced to the size of a grain of sand. Powders were mixed, sintered, dissolved, precipitated, filtered, liquefied, electrolyzed, distilled, separated, and packed into packages. Trolleys brought the packages back to the starting point, wire brushes scraped off the packaging, tearing it into dark shreds, and the exposed bricks were crushed to powder in mortars and thrown back into the chemical process. But it was pure chance and largely thanks to the glass walls that all the production stages could actually be observed. René was the only one to cast a critical eye on the processes. Most of the important things were carried out in secret, inside vibrating and droning pieces of apparatus, whose purpose could only be guessed. Even among the clearly visible processes, much remained incomprehensible: plates were folded irregularly, elastic wires were twisted into strange shapes, balls were inflated, gliding discharges grew up walls into tufts of lichen, threads danced as though on a loom, and strips of netting were rolled around rollers.

Several times Don's group encountered obstacles. As they turned one corner, they came upon a swarm of three-wheeled carts equipped with gripping tongs and spray nozzles, busy as insects, using flying contraptions like little helicopters to build a huge wall. It was already about fifteen meters high and ran right across the street. But that was not enough for these robots – they were tearing down the buildings on the right and left to make room for their wall.

In another street, a chemical factory seemed to be in the grip of madness. A greenish-yellow viscous mass was oozing out of five large openings, leaving only a narrow passage on the opposite side of the street. They hurried through, as the passage gradually began to close up.

"Look at this!" shouted René.

Through the glass wall on the right-hand side, they could see that the bubbly liquid also filled the interior of the room and was already three meters high.

"The walls are bending!" Al shouted. "Run as fast as you can!"

There was a plaintive whimper in the air, and a series of crunching sounds in quick succession.

"The walls are bursting!"

René couldn't take his eyes off the glass surface. It was no longer transparent and smooth, but crisscrossed by a spider's web pattern of cracks. Imperceptibly, the cracks grew more and more numerous, until soon they were so close together that the glass had become opaque, and the wall a kind of mealy white. Then the pressure became irresistible and it finally began to bulge out in a liberating, but deadly gasp. It inflated like a rubber ball being pumped up, then slowly disintegrated into millions of tiny plates. For a few seconds, the splinters stuck like scales to a wide, cylindrical bulge that rolled in leisurely fashion into the street, and then they were buried beneath it.

They ran as fast as they could, close to the left wall. They ran as they had never run before, but they did not run fast enough. Their eyes were fixed on the closing passageway through which they were moving, on the bulging masses that were pushing in front of them, heading as fast as they could toward the narrow and steadily shrinking opening.

"Stop," Al cried and slid a few more meters in a desperate attempt to stop. While the front of the building on the left was smooth and unbroken everywhere else, here a flat construction like a trellis led up the wall. They already felt a tough resistance grasping their feet, so they threw themselves desperately onto this potential lifeline, although it did not offer enough room for all of them to climb up at once. They pushed, shoved, slipped, felt their legs sinking into the flowing and rising mass. They stubbornly pulled themselves up with their arms, trailing strands of sticky threads behind them. Don was the first up, heaving himself onto a narrow ledge. Al was standing on one of the rungs halfway up and held out his hand to Katja. Her face was gray, like a mask. René was still struggling with the slimy masses that were pulling him down like liquid rubber. He was jerking wildly back and forth, when suddenly he bumped into Katja.

The girl let out a shrill scream. Al saw her face fall swiftly away from him. There was an ugly splash as Katja fell on her back into the sticky pulp. She stretched her arms up towards him, but she was helpless. Slowly, but steadily, she was sinking deeper.

Al began to climb back down. René had also noticed what had happened and was vainly trying to catch hold of Katja's hands. Even Don left his safe place to help. Al leaned far forward, and René, realizing that he would not be able to do anything alone, held him tightly by the collar, allowing him to slip a little deeper. Katja was paddling with her arms. She touched the bubbly mass and her right arm stuck there as though frozen. So far, she had been whimpering softly, but now no sound came from her lips. As if her last strength had faded, her left arm also began to sink into the gently undulating surface. Al saw this and let himself fall forward, as far as René's iron grip would allow, just enough to reach her fingertips. He put all his strength into his muscles and pulled. Kat's body lifted a little from the sluggish mass. Al grabbed with his other hand. He realized right away that the hardest part was now over. He still had to make a huge effort, but now Katja could no longer be taken away from him.

What they pulled up over the scaffolding with their combined strength was a shapeless lump, rather like a blob of green-yellow honey stuck to a spoon. Katja was encased in it like an insect larva. Only her face, left arm, and chest were still free. Although her nose had not gone under and her breathing had not been stopped in that way, she was not in fact breathing. And neither was she moving. Her eyes were half open, with a fixed expression and unseeing. When Al and Don laid her on the ledge, the viscous mass spread out around her. They could feel the sticky stuff on themselves. With every step, they had to struggle to lift their feet.

Don stared at Katja.

"Damn it," he said, "we really needed that. She just gave up."

The ledge ran around the entire building. The three of them moved away from the unpleasant proximity of the swollen glue. They left Kat's body lying there – a piece of complexly assembled matter that had now become worthless.

The street on the opposite side was still free. They found another scaffold built against the wall, just like the one they had used to get up, and they climbed down to the ground.

Through rows of pounding machines, under billowing clouds of smoke and falling dust, chased by discharges that whipped through the streets, threatened by falling masts and scaffolds, they approached the center. Nervously, they moved past the porcelain-white, pear-shaped bodies, from which fountains of white-hot air were now mushrooming upwards.

Finally, they entered the corridors that led into the hill, and the noise of the machines gradually faded into an indistinct murmur. The twilight was like a grave. Everything outside seemed distant and insignificant. They did what they had set out to do – to search for Jak and Heiko – but the driving force was no longer a necessity conditioned by their predicament, only a sense of duty towards themselves.

Without knowing how long they had been wandering through the corridors, they finally came upon their two adversaries. They were standing in a pulpit-like room, high up under the apex of the hill; a room which seemed to be as much dedicated to control as the one below, only here the effects it was directed at were easy to recognize. A glass window ran right around, providing an overview of the entire city. Whenever Jak or Heiko pulled a lever or pressed a button, a building somewhere below would fall apart, flames would shoot up, projectiles would fly, machines would explode, or something similar would happen. For them, it was just an exciting game: flip a switch and get the thrill of seeing what happens!

"Hey," Al called from the door, "you're just pure vandals!"

The two turned around. Heiko waved at them.

"Ah, there you are! We saw you coming! Where are your toy guns today?"

"Why are you destroying the city?" Al asked. "What's the point of that?"

"None," Jak laughed. "But it's insanely funny! Look!" He flipped a switch, a gate opened below, and a rocket shot out and curved steeply upwards.

"Not so fast," said Heiko. He stepped up to Jak and pressed down another lever. He then moved it back and forth like the joystick of a small plane, as

Jak had done. Below, a second rocket darted out of the silvery reflections and the smoke and raced towards the first. With a sudden turn, the latter dodged. The second turned around and flew back. Jak swung his rocket around in a circle and let it crash into the other. A mass of glowing debris descended to the ground.

"What d'you say to that?" Jak asked. "Come on, try it yourselves!"

Don did not seem totally averse, but he restrained himself and asked: "How far have you come? Have you reached the goal?"

Jak sat down on a control table.

"We're almost there," he said. "There's just one little thing missing."

Don looked thoughtfully from one to the other.

"Jak," he said. "You've already lost two men. You're outnumbered. I have a proposal for you: take me into your group! I'll continue with you. Like a kind of exchange for René."

Jak drew air in sharply through his nose. "Aha, you're really going for this one!" He thought for a moment then added: "But you're not so stupid. I'll go for that."

Don had regained some of his usual self-confidence. He turned to Al and René: "Did you hear? It's best if you leave now. You're way too soft. Your nonsense has messed up all my actions so far."

Al looked him up and down, then turned contemptuously away from him and addressed Jak.

"You've got a fine addition to your group here! I wish you a lot of fun with him." He exchanged a few quiet words with René, then turned back to Jak: "Listen, René and I are ready to give up. We concede victory to you, but we have just one condition: you must tell us everything you've discovered so far!"

Jak raised his eyebrows in surprise.

"So, if we do find out what they looked like and display the image in a museum, it'll be named after us?"

"Yes," confirmed Al. "We have no interest in that."

Jak jumped from the table and shook Al's hand.

"Agreed."

"Now tell us!" Al demanded.

"Come with me!" said Jak.

They went back down the corridors, through several domed rooms, following the spiraling staircases downwards. Whenever the opportunity arose, they worked their way down. In the meantime, Jak began to make his report.

"There's not much to say," he said. "Our first camp is over there in the west, on the mountainside. That was what we agreed upon. We then set off, first looking at the modern houses and then the old ones, and finally arrived at the wall. We walked along it until we found the bridge. That was our first disappointment – you saw yourselves what was wrong with it. Then we discovered the armory, and I had the idea of making things a bit more complicated for you – it was only to be expected that you would eventually come to the bridge. Were you very surprised?"

"It worked perfectly," grumbled Don uncomfortably.

Jak continued amused: "Now we were three days ahead of you, but it took a while before we got the hang of the reflection. We then climbed down on a rope. Now we were in the center and again we couldn't go any further. Right from the start, the machines had us in some kind of testing facility. Then they let us go again, and we could look around in peace. The factories are quite strange, but we didn't notice any trace of the inhabitants. Did it occur to you that they might have crawled away to hide from some disaster and then just died?"

He looked around questioningly, but he got no answer. Don had no opinion on the matter, and Al was too eager to learn something new to interrupt Jak's explanations with laborious and probably fruitless explanations.

"Of course, this hill seemed the most interesting to us. And that has been confirmed in a certain sense. It's the control center for the city. But that's not all! We found something else, and I want to show you that right away."

The corridors looked much the same, but the rooms they led to and connected were furnished a little differently to those on the upper floors. While the installations there were almost monotonous, there was an indescribable wealth of different kinds of equipment down here. Some rooms seemed to be simple control rooms, while others looked more like laboratories for quite sophisticated physical, chemical, and biological experiments. Still others were no doubt used as archives, with countless compartments full of tapes, disks, rolls, and the like, destined for documentation purposes.

Al was fascinated by all this, but Jak went on undeterred, ever further into the depths. Finally, they reached a long, low hall, whose ceiling was supported by regularly arranged pillars. It was empty. It seemed to bring to an end the system of corridors leading downwards, as there were none of the usual devices which humans and the presumably human-like inhabitants of this planet would have used to separate or connect rooms – no doors or gates, no window openings, no further corridors. However, there was something conspicuous on the floor, just one thing: a dish-shaped depression, about twenty meters in diameter, and within the hollow, further dish-shaped depressions, encircling each other in a ring, but arranged eccentrically, the enclosed ones always reaching a little deeper than the enclosing ones. If one looked closer, one noticed that the round shapes were tiered, as if everything was composed of small cube-shaped building blocks.

Jak stepped up to the edge and said: "This is it. What do you think of it?"

In contrast to the gray material that covered the rest of the floor and the cladding on the walls, which seemed to line the entire interior of the hill, the building material of the floor and the embedded dish was metal; a reflective, but unusually dark, almost black metal.

"There seems to be something underneath," said René. He scratched around a little, then lay down and put his ear to it. "Nothing."

"We should try to open it," said Don.

"But how?" asked Heiko.

Al placed his hand on the edge of the outermost ring. It seemed cool and surprisingly smooth. Al felt as if his hand were connecting with movements and currents that were alive down there, but he immediately scolded himself for having such an absurd idea. However, he could not prevent his heart from beating more strongly. A strange feeling surged up in him, and an inner voice whispered to him: "In one of the control rooms there must be a way to open the gate. We need to figure out how to operate these devices, we ..."

Don interrupted him rudely: "I have an idea." He waited a moment to give his words the right emphasis. "We'll blow the lid off!"

"With what?" asked Jak, interestedly.

Don winked at him.

"With what? That's the question. But I have the answer. In this city there are rockets, so there must also be explosives – bombs, probably atomic bombs. All we have to do is find them and blow the lock."

"That's the most unreasonable thing I've ever heard," Al exclaimed, and stepped towards Don. Then he felt a hand on his shoulder. Jak was grinning at him.

"I don't think it's stupid at all. I even think it's very good! We'll blow this pot open. Why not?" He nodded happily to himself. "You don't have to join in," he said to Al and René. "You can have fun with the circuits, but don't disturb us. We'll be too busy for that!"

Jak, Don, and Heiko soon disappeared into the room through which they had arrived shortly earlier.

Al and René watched them go in silence.

* * *

After their companions had left, everything went very quiet. The corridors were lifeless and lonely. They would have liked to say something, just to break the silence, but there was not much to say. René wanted to search the laboratories, so they wandered up the sloping paths to the next floor.

"If we want to find anything else, we don't have a great deal of time left," Al said dejectedly.

René tried to comfort him. "Maybe they won't find a bomb."

"We'll just have to see!" Al replied without much hope.

They entered one of the laboratories, and for René everything else faded into the background. He started peering into microscopes, turning flywheels, examining scales, following swinging pointers, and projecting spectra onto grid scales. He wandered, completely absorbed, from one device to another, turning, switching, shifting …

"I'm going to take a look around the archives now," said Al, but he wasn't sure whether René had heard him. He looked into a few rooms and stopped in one where there was something that resembled a total immersion facility. He knew it would be dangerous to use such a device because these procedures interfere with brain processes, and it was not clear how the induced stimuli would affect him in this case. First, he examined the control panels on the chair backs. If they were adapted for the same purpose as those on Earth, then the arrangement and adaptation to the reproduction qualities would not differ too much from the ones he was used to. He turned the screws a little, thought hard, and came up with a hypothesis for the operation which still had to be proven correct, but seemed quite likely to be on target. It was a risk, but he wanted to learn as much as possible as quickly as possible, so it was a risk he had to take. He sat in the chair, grounded his ankle, and put the helmet over his skull. It was strikingly similar to the ones used on Earth and happened to fit him extremely well. Then he put his forearms on the armrests and his hands in front of the control screws. The space for his arms was a bit cramped, but if he pushed his elbows back, they fit quite well into the slightly curved, groove-like supports.

Now he pressed a button down with his left thumb. He had previously determined that he could turn it off again by pressing it twice, so he left his fingertip in contact with the slightly concave surface of the switch button and waited anxiously.

The light dimmed.

Al wanted to look at the strip of light, but he could move neither his head nor his eyes. He was forced to stare straight ahead and saw the contours of the objects blur until everything disappeared into a uniform grayness. His sense of sight was thus affected in the normal way. What about his hearing? He muttered a few numbers and words to himself: "One, two, three, four. Can you hear me? Can you hear me?" He moved his tongue and lips, but he heard nothing. He continued to speak: "One, two, three, four, five, …", and suddenly he wasn't sure whether he was really speaking or just imagining it. There wasn't complete silence. He heard a soft noise, but his attempts to speak were unsuccessful. Suddenly, he was struck by a fear. What if he couldn't move his thumb anymore? He quickly pressed down.

Immediately the light began to glow again and objects formed out of the gray.

"… four, five, six, …"

Al heard numbers and only now realized that he was still counting.

He took off the helmet and stood up. He had reason to be satisfied – nothing untoward had happened to him. On the other hand, the apparatus seemed to have gone dead. Then he laughed nervously – he hadn't inserted a tape! He looked for the turntable, but couldn't find one. However, he noticed a small slot. He looked around for something that could fit into it and came across some thin, flexible, but tough metal plates on a shelf. He picked one of them out at random and inserted it into the slot. Then he sat back down in the chair, put on the helmet, and pressed the button on the left.

After a moment, the light dimmed, then immediately brightened again, becoming brighter and brighter, even much too bright! Green suns circled, then melted away. Al's fingers felt for the buttons and the light receded again – colors separated, stripes trembled, jumped, arranged themselves.

No sound, so switch!

A tremendous roaring – stop, too much!

Another button brought an intrusive smell of hay, and tears welled up in his eyes. He sneezed.

Yet another button and the smell receded, the colors became paler, his weight decreased … there was a rotating handle to regulate the intensity of the experience – that was good: a little below normal … now, he could see, hear, smell, taste, and feel. The sun shone, grass brushed against his feet and rustled softly. Next to him was a small lake. Al crept forward – there the red ball gleamed, half hidden behind a rocky pinnacle. Al had found it first! Now it was important to hit it on the right side! He knew that thousands were watching him on their telescreens, and as he set up the slingshot, he positioned himself to offer them his profile. He briefly took aim and the bolt cut through the air. It hit its target! The ball rose in a straight line, inclined at exactly forty-five degrees to the horizontal – a magnificent volley – and only fell back to earth after a good two hundred meters. There was a roar of applause.

Al switched off and the scene faded. He was once more sitting in the dim archive room, leaning back in his chair. He breathed a sigh of relief. He admired the subtleties of the reproduction, the colorfulness of the impressions. It was something new for him that thoughts and decisions were also conveyed here – once again, it showed the superiority of this technology. But above all, the recordings were accessible to him. His sensory organs and his way of thinking were similar enough to those of the former inhabitants of the planet that he could not only relive their impressions, but also found them understandable! It was clear that they must have been humans or creatures very similar to humans. If he let this data plate run further or inserted others, then he would surely see them. Even better, he would be able to observe them, hear their language, even feel their joys and sorrows. Of course, there could be small differences in the categories of sensation, because the control buttons offered him a certain range of variation, and he could use this to adjust the levels to his stimulus and pain thresholds. But the fact that the optimal values were all within the adjustable range confirmed a high degree of similarity.

He had thus achieved what their expedition had set out to achieve – better even than the rules prescribed. But that was now irrelevant to him.

He pulled the metal plate out of the slot and pushed another one in. Again, he made himself comfortable in the playback chair, now with much more confidence than on the first attempt. This time he was lucky: he saw the beings. A whole group of them formed a semicircle around a machine. Two of them were using fork-like objects to attach wires to eyelets which protruded from a perforated base plate like a sieve – it was obviously an important ceremonial act, because all the others were watching in rapt attention.

They didn't look much different from humans. If Al hadn't been sure that they were creatures on a planet millions of light-years away from Earth, orbiting a foreign sun, he would have taken them for humans. Their skin was perhaps a shade grayer, they were small and frail, their heads appeared strikingly large, and their facial features strangely distorted, but otherwise they looked like humans, and they moved like humans. Although Al had expected this, he was shocked to see it so clearly confirmed that this form of intelligent being could evolve anywhere in space, if only the environmental conditions were the same.

This time the sequence would not let him intervene actively – he felt as if he himself were standing in front of the machine, but as an uninvolved spectator.

The two individuals had by now attached all the wires to the ground and stepped back. One began to speak in an incomprehensible, rather nasal language, and the other answered him. From their gestures and emphasis on certain words, Al believed he was witnessing some kind of solemn declaration with a speech and a response. Since he didn't understand what they were saying, he continued to observe them. Everyone wore tight-fitting brown and olive-green suits and some held flat bags in their hands. They were all watching the two speakers attentively. Finally, one of the two stepped back and the other grabbed a handle protruding from the side wall of the apparatus. A screen tipped forward. There were wires running down from it to the ground, and these now began to stretch vertically. At first, nothing moved, but given how attentive the observers were, Al could tell they were expecting something to appear down on the ground, between the wires. And it wasn't long before something did indeed emerge from the holes: yellow shoots, moving unsteadily, as if they wanted to feel their way, growing somewhat jerkily. When they had reached the size of a hand, one of the people in the field of view reached for the control panel and all the shoots divided. They went on growing, and with another switch adjustment, leaves began to sprout from all the stems. The whole plant was gradually built up, step by step, according to instructions. It was as if its growth had been recorded in fast motion, but that was not the case, as the otherwise normal movements of the people proved. It was not an optical trick, but a botanical experiment – the controlled growth of a plant. The two individuals in the middle now resumed their exchange of words. Al switched off. He sat thoughtfully for a while. He had already achieved a lot, but he wanted more. He wanted the past and above all the future.

* * *

He rummaged for a long time in the compartments of the various rooms. The aim was to find those records among the enormous amount of material that would give him an historical overview. He searched for characters and found dot patterns, similar to Braille, punched into the upper left corners of all the sheets. He informed himself about the content of many dozens of the recorded metal plates, and quickly identified the way things were organized. Now he knew where the records on general principles were stored, where he could inform himself about experiments or laws, and where he should search for more detailed series. He noticed that each room was assigned to a certain subject area, and sighed as he set about taking samples in one room after another. When he realized how much time this was taking, he decided to check on René first.

René was busy trying to keep a ball of glowing liquid metal floating in the air.

"Hello, René," said Al. "I've seen the inhabitants!"

René adjusted two pole pieces. Looking down at the micrometer screws, he replied: "That's nice, Al. What do they look like?"

"Like humans."

"Just as we thought," murmured René. By increasing the magnetic energy, he managed to get the soccer ball-sized blob of liquid metal to rotate. It flattened noticeably at the poles.

"If only I knew what kind of force was involved," said René. "Magnetism alone wouldn't be enough."

"That can't be so difficult," said Al. "Have you found anything else of note?"

"A few laboratories," René replied. "You can't imagine all they experimented with. Not just physics and chemistry, even mathematics. And one laboratory even seems to be devoted to history!"

Al perked up.

"Can you show it to me?"

René tore himself away from his tinkering with some reluctance. He let the liquid ball slide into a container, where it spread out. Then he stood up.

"Come with me, I'll show you." They stepped out into the hallway, René flung open a few doors, looked inside, moved on to the next, and finally announced: "Here it is."

They entered a room that differed from the others primarily in terms of its furnishings. It was a kind of mix between laboratory and museum. In addition to some unidentifiable apparatus, there were three playback chairs with associated containers for the circuits. Embedded in the wall were some glass windows covering display cases which contained realistic models of settlements, villages, and cities. Dots and lines moved around within these, only sporadically in the small places, in whole swarms in the large ones.

"Those must be the inhabitants and their vehicles," Al noted. He reached for a control panel and moved a pointer across the scale.

"Let's see what this does."

Suddenly one of the models changed. The place expanded, new buildings sprouted like mushrooms from the ground, trees and green spaces disappeared, and factories with chimneys, road bridges, and airports took their place.

René moved the pointer again.

Explosions shook the city, houses collapsed, craters yawned. Like a storm, disaster struck the settlement, and like a storm, it passed. The torn earth was soon covered with green, modern houses filled the gaps between the old ones, and traffic peaked, then subsided again.

"Indeed," said Al, "history played out."

He was about to turn away from the demonstration when something held him back: instead of the settlement, a flat crater now yawned, and above it hung a mottled white and gray mushroom cloud.

Al shuddered.

"It seems like we don't have much time," he said. He walked over to the other wall, which was filled with compartments. Perhaps things are arranged here in the same way, he thought. He reached into the compartment where, according to his conjecture, the basic facts should be stored. And since it concerned history, what should these consist of if not an overview from the earliest times to the present?

"Maybe we'll find out what happened to them after all," he said quietly to himself, then pointed to a demonstration chair and turned to René: "Sit down and pay attention!"

He was by now quite familiar with the equipment. He noticed more and more details that he also recognized from Earth. He connected the two chairs with a cable and set them to receive the same presentation. Then he inserted a recorded plate into the slot and placed others on a table, from where they would slide down and gradually take their place in the scanner once the first had been played.

Al and René put on the receiver helmets.

What happened was exactly what he had expected. The presentation began with a small group of rather primitive-looking fur-clad individuals creeping through dense forests armed with stone hand axes. They had seen such images before, and the pictures themselves did not much impress them. But they could also hear the cries and shouts with which these Stone Age people communicated. They could smell their sweat and blood, and feel vermin on their hands and thorns in their bare soles. The impressive thing was that they did not feel all this as they were used to feeling it, but rather as these dull, primordial beings themselves would have felt it; and indeed, they also experienced the mental lives of those beings. There was nothing clear and pure about these impressions, although everything was there, and it was completely there, from the water they slurped, the trees behind which they hid, the rocks under which they foraged, the flesh they ate, the females they desired. Everything was real, but it seemed strangely vague, strangely veiled. They experienced feelings that they themselves were used to experiencing: joys and fears, anger and submission. But some of them were disturbingly strong, with no possibility of control, with no ability to do anything about them – immediate and fateful feelings.

Al shook off the spell as vigorously as he could and pressed the accelerator. Images flickered and faded, and new ones emerged. Some he felt he recognized and yet at the same time they were novel. And feelings arose that seemed both familiar and strangely unfamiliar.

Middle Ages. Castles. Armor. Charcoal fires and wrought iron grids. Superstition. Cults. Persecution and torture. Hate and fear. Dull entanglement in childish notions. Uncertain hopes for reward in the afterlife. Dirt. Disease. Pomp and slavery. Atrocities and remorse.

Atomic age. Barracks. Traffic accidents. Marching columns. Bombs. Delusions of grandeur and powerlessness. Recklessness. Lies. Oppression. Thirst for knowledge. Fear of unleashed natural forces. Arrogance and overestimation. Clinging to outdated traditions. Emotions and instinctive actions. Distrust and excess. Intellectual failure. Mass distress and mass death.

Pacifist state. Garden cities. Terraced houses. Projection screens. Demonstration chairs. Machines. Hovercrafts. Security. Contentment. Fulfillment. Home culture. Art. Tranquility. Games. Illusions. Health and pleasure. Satiation and boredom. Freedom from responsibility. Intoxication, dreams, sleep …

Al went back a bit and slowed the process down. Beyond a certain point, he wanted to know everything in detail. The flickering images merged again into a continuous sequence, the rising and falling noises into articulated sounds, the mood fragments into logically understandable emotional fluctuations.

He saw the city, the old fortress on the hill, the new castle, surrounded by its moat, the medieval wall, the huge modern metropolis which grew right up to the mountains. Then there was a bombardment and only a huge pile of rubble remained. Slowly, buildings emerged from it, more uniform and modern, then others, even more modern; generation after generation of buildings – until finally, the encircling garden city was created. Around it stretched a lovely landscape of meadows, lakes, and rocky outcrops – planted meadows, artificial lakes, and manufactured rocks. The historical, circular inner city was restored, and the old castle was created in the center as a lighting backdrop. Underneath all this were the machines. They took care of everything and one saw as little as possible of them.

Then the planet was struck by a meteor swarm and they built the invisible shield over the city. Despite the regained security, they rarely left their houses. Most of the time they sat in front of the projection screens, looking out into the action, letting themselves be fooled by fairy tales, blending into what is going on or participating only passively in it. They no longer needed to move – everything they wanted to experience was suggested to them by the total reproduction facilities. They didn't need to eat anymore – pipes flushed liquid food to their seats. And so, they just sat, dreamed, or slept …

for days …

for months …

for generations …

While this was happening, Al's and René's empathy grew steadily. They saw things more clearly, they experienced the sensations more clearly, and they understood better what was happening. At the stage they had now reached, veils began to fall over the scenes once more, and their emotional reactions were seized by a strange paralysis. The colors became paler, the outlines blurred, everything took place in mysterious semi-darkness, often dissolving into incomprehensible impressions, but in fact, very pleasant impressions. Noises sounded like music, body feelings disappeared, scents swelled up intoxicatingly. Then, from this rather formless backdrop, clearer impressions suddenly emerged again: someone transported in a hovercraft climbed two hundred meters up in the air, interrupted the connection with the energy rail, and came crashing down.

Again, there was a half-darkness that they did not understand. Unrecognizable things slid by – shadows on a magically lit stage. Even wishing and wanting gave way to something else that had goals without striving for them, intentions without realizing them. Al turned the knobs, but achieved no greater clarity. What was described here followed different lines of thought to human brains.

And now, Al's hands clenched the supports. From the shadows emerged an even darker blackness, a dish-like metal hollow, the gateway to the innermost, yet open … into the wide hole sank an endless series of cylinders. Gray spots passed across the scene. A square pattern went by, dotted with points of light, and pale stripes shone, liquid foamed, wires forked and reunited, pointers moved, contacts closed, and a violet light hung like steam in a corridor. The air was moist and warm, and cables, wires, tubes, reflectors, and lamps formed enclosed areas reminiscent of large birdcages. An endless row of these was arranged along the corridor, and in each sat a pink, fleshy, many-lobed structure. Each sat in a bowl filled with liquid and each was supported by stays like a precious and delicate plant. Each outgrowth was enclosed in a casing, with wires and tubes protruding into it – transparent tubes with colorless, yellow, and red liquids pulsing through them. Each looked like an orchid locked up in a cage.

Here the presentation broke off.

Confused and dazed, Al and René rose from their chairs.

"Did you understand anything?" René asked.

"The history of the city is clear to me," Al replied. "It's a perfectly normal story, just like what could have happened in any of our cities. They also went through their atomic age. The bombs that destroyed the city were undoubtedly atomic bombs. I'm quite sure that some of them are still stored here somewhere."

"Let's see if Jak has already found them," René suggested.

As they searched for the way up, Al continued: "We now know the explanation for the invisible shield that initially gave us such a headache."

"A meteor swarm," René remembered. "But why didn't they remove the craters from their landscape? With the means they had at their disposal, that should have been easy for them."

"Apparently, the traces were even easier to cover up in another way. These transparent walls that we found in the houses are light screens. Absolutely anything can be projected on them – near and far, past and present, real and imaginary. With a small change in the optical controls, no one saw anything of the ugly impacts anymore."

"And the purpose of the mirage was also to create just such an illusion. How could anyone be satisfied with something like that?!"

"Why not? Remember how much of our own world is deception, sham, and forgery!"

"Yes, true! But at least we know when something is fake!"

"Does that make it any more real?"

René didn't answer. They had reached the upper observation tower and looked out. The whole area was a picture of devastation. The air was no longer as clear as before and they struggled to find Jak, Don, and Heiko in the chaos.

"Do they know anything about technical matters?" Al asked.

"Heiko knows quite a lot," said René.

This was soon confirmed. Heiko disappeared round a corner, then immediately reappeared on the high seat of a tractor. He was directing an articulated pincer arm through an open door. Jak and Don squeezed past it into the room. After two minutes, the arm moved again. The sections of the arm folded together. A heavy metal sled hung from the bars; on it a rocket projectile in its launch device. They could see it clearly, stuck in the cylindrical gun carriage.

"They're heading for the outskirts of the city!" said René.

"Maybe we can still prevent it!" Al shouted. He tried to commit to memory the layout of the buildings and the location of Jak's team.

They rushed down the sloping tracks and through the dark corridors to the exit. The light was blinding, but they ran on regardless, around piles of rubble, collapsed buildings, and crater holes. Much was destroyed, but much also remained untouched. Al's gaze passed over this perfect construction as if he were seeing it for the last time.

And indeed, it was the last time he saw it.

They reached the others just as they arrived at the city wall. "Wait!" Al shouted from afar. "We've found films of the inhabitants. You can beam the images to Earth!"

"What do they look like?" Jak asked.

Al stopped, panting.

"Look at them yourself! They're fantastic shots!"

He could tell Jak was in doubt. He looked at the gleaming body of the missile, which lay like a fish in a net, and then back at Al.

"Don't jeopardize this opportunity!" Al warned. He tried to take advantage of Jak's hesitation.

"Are you going to accept?" Don asked. "Are you so easily persuaded?"

"We have time," said Jak. "We could take a look at it!"

Don was suddenly boiling over with anger again.

"Who knows what will happen then! The machines could come at any moment and take everything away!"

Heiko intervened.

"I think we've disabled the communications network in the control center. So, what could …"

"You're just cowards," said Don. "Cowardly and stupid at that. Well, go on!" he suddenly shouted, "I'll wait here in the meantime and play around with this thing! Maybe I can get it to start!"

"Just be careful!" Jak warned. And to Al: "Did you find out what was under the lid?"

"We didn't have enough time," René explained. "But we also saw something of the lower rooms."

"What?"

"Hard to describe," answered René hesitantly. "There are … well, things like flowers. They look like orchids. And they're in cages."

Don laughed out loud, but didn't comment.

Jak furrowed his brow.

"Is that all?" he asked. "Did you see any kind of creatures?"

René felt cornered.

"Just the flowers."

"An orchid cage. I see," said Jak. "Well, then it's clear what we have to do. Heiko, get the rocket ready!"

Heiko was still perched up in the narrow driver's seat of the tractor. Now he started it up to bring the rocket into the right position.

"You've been duped," Jak said privately to Al.

"Unhook the latch!" Heiko ordered. Don jumped across and released the bracket. Heiko drove a few meters off, jumped down, and came back over to the sled on which the scaffold with the rocket was mounted. He adjusted it and aimed at the hill.

"You're going to blow everything up!" said Al. "Nothing of what we could still find will be left."

Don acted as if he hadn't heard. He stepped up to the scaffold, reached through the bars, and slapped his hand on the metal.

"It's a bit on the small side, this thing," he said. "Is this supposed to be an atomic bomb?"

Heiko took him seriously.

"It's enough for our purposes, you'll see!"

"Can we fire it now?" asked Jak.

"Are you really going to stay here when you shoot?" asked René, horrified.

"What are you afraid of? The radioactive cloud?" Don mocked.

"Jak," said Al, emphatically, "René is absolutely right. We'd be much too close to the explosion. You'll blow us all up!"

"We can't go any further back," said Heiko. "The wall is just behind us. How am I supposed to get over that with the rocket and launcher?"

"That's right. We can't get any further back," Jak confirmed. "We have to fire it from here. If you want, you can run away. But hurry up, we can't wait very long!"

Al shook his head in disbelief.

"You have no idea how powerful this bomb could be. It may not even be an atomic bomb!"

Well, what else could it be?" Jak asked.

"It might be something much worse," Al answered, staying perfectly calm, as if he was dealing with undisciplined students. "Their technology is way ahead of ours. Think about the shielding! They could have developed weapons that we can't even imagine. Are you really going to risk everything …"

Don couldn't bring himself to listen to any more.

"Can't you just stop preaching!"

"Let him go on," said Jak. "What can't we risk, Al?"

"Everything there is here! This opportunity to learn something new. But not only that! Jak, these people have come much further than us. They've survived an atomic age and it's the first time we've come across such a culture – except for ourselves! But where are they? What happened to them? I need to know that, Jak, please understand. Here we can learn what will become of us one day!"

Jak looked thoughtfully at him.

"Alright, Al," he said. "I let you talk. I listened to you. You want to know what happened to them. Fine. I don't understand why that's so important, but that's your business. Now listen to me. I also want to know what's down there in that hill, and soon. Soon or not at all. I'm saying this really clearly to be sure you understand me." Now his voice became sharp: "Because I've had enough of being here. It's so boring. I'm disgusted with it! I just want to have a go at lifting the lid and look inside, and then: goodbye! If it doesn't work, that's fine too. I don't care what happens to us. And even less what happens to the city! Let it blow up. And that's why," he became quiet again, "that's why we're shooting now."

Al nodded. There was really nothing to be done. He watched in a daze as Heiko peered through the sights again and then looked expectantly at Jak. The latter raised and lowered his arm and Heiko grabbed a trigger hanging from a long lead and pressed the button.

The previously lifeless metal body of the projectile began to tremble, and the whole structure trembled with it. Then a fiery jet shot out at the rear, the rocket moved forward a meter, hissing loudly. For a moment, it seemed to come to a halt once more, and then it swept away at tremendous speed, making a beeline for the hill.

There was almost a second during which nothing happened. Then all of a sudden, the air seemed to tear apart. It all happened in a stunning silence. The last thing Al saw was a wall of fire rushing towards him.

The Third Attempt

Up here the sun burned much more strongly than it had once done down in the valley, but the wind was also stronger and cooled all the shaded areas to freezing point in a matter of seconds. Sometimes, it brought clouds of sand, fine sand that soon got into their eyes, ears, and mouth, grinding unpleasantly between their teeth, filling their clothes and chafing their skin whenever they moved their limbs.

Al and René were standing on glassy crusts and slag. They looked down at the plain, a hundred meters below them. The air was still saturated with that intoxicating scent of thyme. Its origin seemed even more mysterious than before, now that there were no longer any plants.

This was the third time they had awakened on this planet. This time they had to wait longer, for nothing of the facilities had been preserved. They did not know whether the explosion had destroyed everything or whether the transformation had occurred during those inexplicable events that must have occurred afterwards.

The city was no longer. There was no plain and no valley. There was only a desert that lay like a sea in the basin below them, a good two kilometers above the level of the old valley floor. Much of the surface was covered with sand, but it could not have been very deep, for in some places, rocks protruded above the surface. These were not cliffs or boulders, but bulbous masses, sometimes also flat steps. It looked as if someone had poured out wax and it had solidified before its surface had had time to completely smooth out. This desert extended without interruption to the opposite mountain slopes, which stood as a dark line on the horizon.

"I think there's no point anymore," said René.

© The Author(s), under exclusive license to Springer Nature Switzerland AG 2024
H. W. Franke, *The Orchid Cage*, Science and Fiction,
https://doi.org/10.1007/978-3-031-60499-7_3

"Do you think everything has been destroyed?" asked Al.

"At least nothing is left of the city."

Al looked at him questioningly. "Have you thought about the mirages?"

René laughed in surprise. "That would of course be an explanation for everything. Why didn't I think of that! Do you think the stones and sand are all illusions?"

"We'll soon find out."

They had set up camp at a spot from which they could easily descend through a valley. Just above it, the walls began to rise steeply. To gain a flat surface for the buildings and as a landing place for the helicopter, they had made a wedge-shaped cut in the rock.

Even at this height, the traces of the shock wave were still visible. Protruding rock peaks were rounded by sintering, and now hardened ridges of once molten rock made a pattern like the tendrils of climbing plants. As Al and René climbed down, thin layers of solidified material were splintering off under their feet. Sometimes they sank into holes concealed by the fact that they were filled with sand.

"The ground is real enough," René concluded when he reached the bottom. He bent down, picked up a handful of sand, and let the soft mass run through his fingers. Al waded a few meters out and began to brush it aside with his feet. After a while, he called René over to him.

"What do you think of this?"

René knelt down and felt the smooth surface that had appeared under the layer of sand.

"Plastic," he said. "The same plastic substance used to make the rocky pinnacles in the lake and meadow landscape."

"I think it makes sense," said Al. René looked up at him, puzzled.

"I think it makes sense to look more closely around here. There is something else here that is still able to cause change. There's nothing natural about these plastics."

"Now I understand," said René. He stood up again and dusted his hands on his pants. "They're still here. They're alive, and they've filled the valley with this. But why?"

"Maybe because they want to protect something that lies beneath."

"An incredible achievement," René said. "And all in just fourteen days! We've missed a lot!"

They slowly trudged back to the mountainside.

"This sand is awful," René complained. And then something occurred to him. "Hey, Al! Where did it come from?"

"Easy to guess!" Al smiled amusedly. "It's radioactive ash that sank back down from the atomic mushroom after the explosion."

René was shocked to hear this. He ran ahead as fast as he could and jumped onto the rocky ground.

Al followed at a leisurely pace.

"You're no safer there. I'd guess the whole surface is radioactive."

He watched René with amusement. Then he said, "What can radiation do to us? Don't forget, there are no rules anymore!"

René breathed a sigh of relief.

"That's so strange," he said. "I still haven't got used to it."

"It's just as strange for me," said Al. He had caught up with René, and they climbed up the slope side by side. "But the thing about the radioactivity is the easiest. You don't feel it anyway. We can just ignore it. But strictly speaking, nothing forces us to keep our usual sensory impressions. For example, what harm can cold do to us? If it bothers you, you can simply turn it off! I wouldn't recommend that in hot weather, though."

René was confused, but he didn't want to show it.

"Sure, heat would be harmful. But you could raise the pain threshold quite a bit. And what about vision? Wouldn't it be appropriate to increase the visible spectral range? For example, beyond ultraviolet?"

"You could do that. I don't think it would help you much, though."

"If we no longer need to follow the rules, why don't we use more powerful models? Ones that allow us to hear and see better?"

"There are no longer any others." The old ones from the rocket era are no longer operational. Anyway, the finesse in their implementation was rather limited, even though the range of reception was broader. We would have had to develop something new and that would have taken too long. Maybe we'll still have to do that anyway.

"But there is a second reason. Such a model captures very different qualities than we are used to with our own sensory organs. How long do you think it would take for the human brain to learn? As long as we only have familiar impressions to process, we can react quickly and safely. And I think we're going to need that."

They had returned to their artificial platform. The land below them seemed indescribably empty and desolate. Now that they knew that something was perhaps hidden somewhere below, something whose motivations and intentions they could not understand, it all felt a whole lot more threatening.

Al had been silent for a few minutes. The wind pressed the clothes against his body. He shivered and pulled up his collar.

"I'm cold," he said, "but it's strange. I feel comfortable with that. As long as it's not necessary, I'll leave everything as it is."

"I feel the same way," said René. "It seems wonderful to me to have a real job to do. To really be able to achieve something. To face a serious opponent."

"We'll have to get used to it first," said Al. "Actually, it's an incredible coincidence that we stumbled upon something here that differs from everything that has been found so far."

"Maybe others have already found something similar, but just ignored it? I mean, maybe they just gave up like Don, Jak, and Heiko?"

Al felt something strange. He suddenly had the impression of not living in a self-evident world anymore, but in the midst of a web of enticing secrets and puzzles.

"Couldn't it be …," he said. "I mean, couldn't there be much, much more that we just don't know about in space? Lots of things that are really worth finding out about, if you put your mind to it?"

René could not answer this question for him, but for the first time he felt he understood his companion's sometimes strange trains of thought.

* * *

The helicopter carried them over the radioactive desert. The wind shook them, lifting them up and then letting them fall. They swayed back and forth, and just as in similar situations back on Earth, it seemed as if an underground force was lifting up the landscapes beneath them.

"Are you expecting a protective shield – similar to the one over the city?" asked René.

"Yes," said Al. He stared into the shimmering void before him.

"What then?" asked René.

"Then we'll have to carry the equipment."

Contrary to their expectations, they continued unchallenged. Nothing held them back. There were no illusions to trick them.

"The center must be somewhere here," said René.

Al pulled on the joystick.

"I'm going down."

He steered towards a flat rock surface. From small holes, fountains of sand whirled up. He landed gently, opened the door, and jumped to the ground. He sniffed. Strangely enough, it smelled of thyme here too.

René handed Al the explosives and the support frame for the seismograph. Al took the adhesive capsules and carried them about twenty meters across the rock slab. Then he secured them to the ground and pulled the detonation wire back to the aircraft. He connected the end to one pole of the battery via a switch insert and grounded the other.

René adjusted the seismograph and set it going as a test. A slightly wavy line appeared on the output tape. He fiddled nervously with the device.

"What's up?" Al asked.

"The background noise is much too weak," René explained.

"What does that mean?"

"There are always vibrations running through the ground. The device picks them up. That's why the zero line is wavy. But the deviations here are much weaker than they should be."

"Let's give it a try," Al suggested. "Are you ready?"

"Yes."

Al pulled the switch. A small fountain of rock and dust sprayed up from where the explosive had been, and the sound of the powerful explosion rang out.

Their eyes were fixed on the punch tape that was snaking out of the slot of the seismograph. Not two seconds passed before the pointer swung and the nozzle sprayed several sharp spikes onto the paper. René was just about to sit up, satisfied with the result, when there was another rumble and the sound of thunder around them. Since it had been completely quiet in the meantime, it now seemed twice as loud.

"An echo," said Al.

René looked at him, shaking his head.

"Maybe, but from where?"

"Probably from the mountains," Al speculated.

"It wasn't from the mountains," said René. "It came much too quickly."

Al looked around in surprise.

"You won't find anything nearby that could produce such a strong echo," said René. "And besides ... it seemed to me as if it came from above."

"Oh," Al said, puzzled.

"Prepare another charge," René requested. "We need to find out!"

Al complied with his request and ignited the explosive. They tilted their heads to better determine the direction.

The capsule detonated with a loud bang. There were seven seconds of silence and then came the rumbling of rebounding sound waves.

"Good heavens!" Al groaned. "It really does come from above!"

René furrowed his brow in deep thought.

"It can only be one thing," he exclaimed. "The invisible shield!"

"Holy smoke! I think you're right!" Al was full of admiration. "Of course, the shield! They've raised it!"

"But why?" René asked.

"They wanted to complete the protection!"

"That means they don't know how we got here."

"Exactly," Al confirmed. "The synchronous beam goes through the shield, because we had no reception problems when we were under it."

"They don't know that," said René. "We've outsmarted them. In some ways, we're superior to them. I must say, that boosts my confidence!"

Both were excited, as if they had won a victory. In good spirits, they turned back to the seismograph.

"What do you deduce from the curve?" Al asked.

"One thing is certain: down below, about two kilometers down, there is a reflective layer ..."

Al interrupted him: "Maybe the ceiling of the basement rooms?"

"Could be. Anyway, I think I can now explain the weak background signal. Between this layer and the plastic surface, there must be a highly damping material."

"Excellent!" exclaimed Al. "Then everything is clear! For the first time I think I understand what's going on here. It's all about protecting what lies beneath. Apparently, the deepest rooms, the ones we couldn't reach, are still preserved. They contain something valuable. The atomic bomb explosion has proven that the shield over the city and the optical illusions are not sufficient to preserve it, so more effective measures have now been taken. The shield now extends much further, high into the mountains, perhaps even beyond ..."

"Perhaps even around the entire planet!" added René.

Al agreed.

"That would also be conceivable. Besides the shield, they've also installed something else, a thick slab which lies directly above the basement rooms. It's made of a damping material and it's designed to absorb shocks! That's almost certainly it!"

René fully agreed with Al's conclusions.

"That could well be true. The depth of the reflecting layer should also be the same as before."

"How deep is it?"

"I can't say exactly, because I don't know the speed of sound in this damping material, but as I said, it must be about the depth of the old valley floor, at the level where we found the strange entrance under the hill."

René cut off the perforated and rasterized strip containing the seismogram, wound it up, and put it in a small box on the side wall of the device. Then he closed the lid.

"The big question is: How do we get down there?" he said and slung the carrying strap over his shoulder. As he did so, his gaze fell to the west and he suddenly froze. A shadow was racing across the ground; a dark spot, weightlessly gliding up and down the bumps, darting over flat strips, and leaping over pits – straight towards him. Immediately, he lifted his eyes to look for the cause, but the sun was shining directly in his face and he could only make out one thing clearly enough to be sure of it: there was a dark body of indeterminate size in the shape of a hanging bell. He just managed to let out a scream before the shadow was over him and then saw nothing more.

Al only realized that something had happened when he heard the scream. He saw how the bell was placed over René, and ran towards the helicopter. But before he reached it, a shadow seized him too. A black maw opened above him and quickly lowered into position, closing up under his feet. He felt himself lifted up a short distance, and it became pitch dark.

He stretched out his hands tentatively and tried to walk over to the wall. He took steps, but never quite reached the wall. He had the feeling that the ground under him was mysteriously adapting to the movements of his feet, that it was somehow compensating for them. He stood still for a moment and, leaning forward, tried to touch the ground. What he found was something solid, yet flexible, like a board sitting on springs and attached with hinges. But he knew that this comparison was far too simplistic and that the reality went way beyond his imagination.

Suddenly, something stirred, a light glowed for a brief instant, and a sound rang out, but was already swallowed up before he really became aware of it. Something touched him and he felt a small pain, so briefly that he could not be sure whether he was mistaken.

This event happened around him in a flash, yet carefully, even gently, but with unswerving insistence, without restricting his freedom of movement in the slightest, without offering him the slightest opportunity to rebel.

It's a test, he thought, a test like the one a little more than two weeks ago when we first entered the machine area. Everyone who crosses the border is tested. This bell, however, was quite different from the large room where they had been transported from cell to cell and treated in a relatively rough manner. Nothing unpleasant, distressing, or frightening happened here – there was something almost perfect about it. The similarity of the two processes was unmistakable, but the primitive circumstances of the first had now given way to an incredible technical superiority. It was as if the method had evolved from its beginnings to perfection in just two weeks. However, Al realized that this was not possible, that it could not be so. This highly evolved thing must already have been there. It was just that it had not intervened. It had left the security tasks to simpler, automatic mechanisms. But presumably, these no longer existed, and now he was in the hands of something against which he was even less able to rebel than the tests in the machine city. Back then, they had been found to be harmless, but in the meantime, they had shown that this was not actually the case. The old machinery had failed. The test result had been wrong. Would the new one also be mistaken? And if not, what would happen to them then?

The decision was made. Al didn't have to wait long for it, he just didn't know what it turned out to be. He sank a meter deeper, and found he stood on solid ground again … and from the ground, a dazzlingly bright horizon grew up. The bell lifted and set him free. A shadow darted away, the massive metal body melting into a point somewhere in the distance.

"Hey, Al, are you still alive?"

Al turned around. René was standing behind him. He was in the exact spot where the bell had caught him, and Al had not been moved a meter from his position either. But where the helicopter had stood, there was now another bell, much larger than the one that had covered René or himself. It was made of that shiny black metal that they already knew from the gate to the underworld, from the saucer-shaped slab at the base of the hill. Al was about to approach it, when the towering body also lifted with the same ease as the two smaller specimens and shot away.

"I've had enough of these surprises," grumbled René.

"Well, we rather provoked them with our test explosions," Al said. "Maybe they're allergic to explosions. I'm more interested in what kind of results these investigations will give. Even the helicopter was tested."

"Apparently none – they seem to be peaceful. At least, they've left us in peace."

"I'd be surprised if everything went as smoothly this time," sighed Al.

They looked suspiciously out over the empty plain. In fact, they were looking precisely at something strange that had just started happening. The sand had begun to bulge as if some kind of creature was trying to stretch itself up underneath, and then a black cylinder emerged. It grew until it stood like a small, squat tower lost in the desert.

* * *

In just a few minutes, the mood of the two of them had changed fundamentally. Before the appearance of the test bell, they had felt as if they had already achieved success; as if there were only a few minor technical obstacles to overcome to allow them to reach their goal. And now the other side had taken the initiative again, and in a way that made their motives as incomprehensible as ever before.

"What does this mean?" asked René.

Al thought for a moment.

"They've examined us and reached a conclusion. That's the answer to that."

"Do you think this black tower is there for us?"

"In a way." Al made a decision. "Let's take a closer look at it!"

René was not thrilled with the suggestion.

"It could be a trap."

"This tower seems to offer us a way to get down and that's exactly what we want. If they intended to capture us, kidnap us, or do something else to us, what could we do about it anyway? Or would you be able to resist these bells?" He waited briefly for an answer, but René remained silent. "Well then," Al continued. "This seems more like a peaceful invitation to me. I'm going to follow it anyway."

"How can you even judge whether they mean well or ill? You admit yourself that they think quite differently to us."

"I do indeed. But do you think it makes sense under these circumstances to try to penetrate the world below in a different way? Do you want to make a tunnel or dig out the whole area? Do you think that would be safer?"

"Alright," René said after a while. "Let's go then."

As long as they were on solid ground, they made good progress. However, they also had to deal with the sand. They sank into it as one would sink into powdery snow. Fortunately, the ground was never more than a few decimeters down, so they were able to approach the tower slowly, and without being significantly hindered.

This structure too was made of the black shiny alloy, like everything they had so far encountered from the area of the lowest regions. It seemed to have emerged directly from the sand – when René got too close, he slid into a sand-filled crevice between the wall and the rocky base. If he hadn't braced himself with his knee, he would have sunk even deeper. Al held out his hand and pulled him back up.

"Phew," gasped René, startled. "There's a shaft leading down here."

Al winked with good-natured mockery.

"If we want to go down, I suggest we do it inside the tower, not next to it." He pointed to an opening they hadn't noticed before – a rectangle about one and a half meters high and three meters wide cut out of the curved wall.

"Fine by me," said René, resigned to his fate.

Al stepped in front of the door and stood in amazement.

"Oh, how considerate!"

A ramp had unfolded from the threshold. The similarity with the ramp that had led to the hovercraft was unmistakable.

"I think these machines are sharpening up the knife before they cut us to pieces," René said half-jokingly.

They walked side by side into the room, stooping slightly. It was shaped like a circular vault and it was empty. Across the ceiling ran a line of luminous discs, which René noticed were attached to a row of black blocks.

He was still engrossed in these when Al nudged him: the door was closing. The daylight faded, and only the soft white of the light circles illuminated the room. Then the floor gave way beneath their feet. They sank into the depths.

"An elevator," said René.

"Would you rather use the stairs?" Al asked.

It took a long time. They could tell how fast they were going by the extent to which they seemed to get lighter. Still, it seemed to take an unusually long time before they felt increasing pressure from below, a sign that the elevator was coming to a stop. Then they had the impression that they were moving upwards again, and only realized that it was an illusion when the sliding door opened. Apparently, they were located directly above the slab that separated the basement of the hill from the upper levels. They stepped out and saw that the shaft they had traveled down was directly above the oval opening. They stepped out onto a pedestal, the surface of which connected directly to the floor of the elevator cabin and from which they reached the slab four meters below via a spiral ramp. Only then did they see the gate. The innermost cover had swung upwards – the entrance was open to them.

"We're falling more and more into the power of someone else's will," said René.

"All we can do is hope it's not a destructive one," said Al. "It's too late to turn back."

Prepared for anything now, they moved across the bridge, down to the ground. The hall looked a little different than on their first visit. The ceiling seemed to have been built anew. While it had previously been made of gray material, it now sported a light yellow and brown marbled effect. And there were so many pillars holding up the ceiling that they could hardly see more than a few meters ahead.

"Something new again," said René.

"The pillars?" asked Al, who had been following René's admiring gaze.

"Their arrangement. It is completely irregular. A static distribution. I can think of a whole bunch of reasons off the top of my head why this arrangement provides more safety against collapse than any regular pattern. But I'd have to do the math."

"Yes, but not now," Al cautioned with a touch of irony. He looked into the opening. "I'll take the risk!" He sat on the edge and let his legs dangle. Carefully he leaned forward, felt for a hold with his toes and gradually shifted his weight onto it. There was no floor here, and Al had to concentrate his whole attention to reach a spot four meters below from this very narrow horizontal standing position, a spot where he could at least hold on without balancing like a tightrope walker.

He looked back at René, who preferred to move on all fours. It felt like climbing across a circus dome from one hanging floor to another. It looked dangerous, but that was not what had suddenly given Al a feeling of panic – the lid had closed silently. They were trapped.

When René had found secure supports for his buttocks, elbows, and feet, he noticed the reason for Al's dismay and had to struggle for composure himself.

"The lid's closed. Does that surprise you?" he asked.

"Actually, it's not surprising. But I didn't think about it."

Up to then, they had still been mentally in their usual environment, an alien one to be sure, but one in which every detail could be measured by human standards. They had climbed down a few meters and had only had to grasp the unnaturalness of their immediate surroundings. But now that all means of retreat had been closed off, they realized not only externally, but also internally, that they were in a very different world.

They were hanging in a kind of scaffolding that was built up from only one type of component – the same blocks that had already caught their attention in the lighting fixture of the elevator cabin. But these blocks seemed to be about much more than just lamps. Each block had the exact shape of a cube. The side surfaces were incomparably smooth, although they were by no means without structure. On the contrary, in addition to the luminous discs, there were several dark spots that were just as precisely ground into the surface, and lines that ran straight, parallel to the edges, or in concentric circles around the circular disks.

René ran his finger over the ice-smooth surface of the cube and paused.

"What do you think of this?" he said.

Al was ignoring the details and trying to get an overview. The whole place was made of these cubes, which were lined up like bricks, but not in the form of walls; rather as multi-angled scaffolds. The way they were attached to each other seemed so strong that gravity had no influence on the arrangement. Rows of adjacent cubes formed long beams, from which side struts extended, sometimes leading to other cube towers, sometimes ending freely, and often carrying whole clumps of regularly layered building blocks. There were no walls and no floor – the space extended in all three dimensions. Disks shone on the free sides of the building blocks, each emitting only a weak light; but taken together, they provided a completely uniform, non-spatial, shadowless, dull brightness, which hung milky and cloudy in the free spaces.

But this was not all still and dead. A peculiar movement seemed to be animating the scaffold structure, and sometimes closer, sometimes further away, it chirped, puffed, buzzed, gurgled, crackled, and cracked.

Only then did Al follow René's suggestion and touch one of the cube surfaces facing him. He immediately noticed what his companion was trying to tell him.

"This pattern has a meaning," he said. "There's is a warm spot here … and it's vibrating."

"They're like organs – here, whatever it is emits light, and also sound, heat, and who knows what else."

"… and probably also detects," Al added. "Do you think there are transmitters built in to relay the information?"

"I think information is being passed from one cube to another." He examined the surface pattern again and tapped on two brighter points. "These could be the contact points."

"So, we're under surveillance," Al said. "Compared to this, the masts with the spherical objects were harmless." He reached for a cube and tugged at it. "It seems incredibly firm. It doesn't move at all."

"At least we don't have to worry about the whole thing collapsing under us."

Al turned away from the cubes.

"How about doing a bit of climbing?"

"If we must!"

Al scouted for a path like a mountaineer planning a difficult route up the wall, and then climbed down. It went surprisingly well, and they soon settled into a satisfying steady pace they were able to maintain.

"Do you think it goes on like this forever?" René asked after a while.

"No," Al replied. "There must be something else somewhere. For example, the control center."

"Like a brain? I'm afraid that's too human a thought. Why should the thinking functions be concentrated in one place? Because it's common in organic beings? Now that we've already interconnected all electronic brains over great distances on Earth, that's long outdated. And especially here? It seems to me more like each of these cubes is an equal part of the whole. We could go on looking for a long time."

"Yes, that is probably correct. But there must still be something else here. I believe this system emerged from the machine city. Its purpose cannot be an end in itself. It must achieve some task. And I am convinced that it does indeed achieve its task perfectly."

René was doing gymnastics on a perforated piece of wall that looked like a misshapen ladder. With every move he made, the countless points of light seemed to regroup into new rows and patterns. Sometimes, it looked as if the whole room consisted of nothing but floating lights.

"What kind of task could it be?" asked René, who had reached a horizontal surface where he could rest a little. "The actual task of such machines couldn't be to look after people. If you thought you were going to find some sign of the last inhabitants of this planet, then it looks like you were mistaken – you must surely realize that now. Because there's no trace of any people, and this arrangement is not adapted to people in any way whatsoever. So, what is the task supposed to be?"

Al had also sat down. He was sweating profusely from the effort of climbing.

"What task?" he repeated. "I admit that there are contradictions here. But I'm convinced that everything can be explained logically. We just haven't thought the situation through thoroughly enough. We need some kind of clue that can put us on the right track."

"But we don't even know what to look for?"

Al whistled quietly to himself for a while. Then he said: "There is something we haven't found yet: the flower-like structures in the long corridor. The orchid cage. There must be some good reason why the film sequence in the history laboratory stopped just at that point. Maybe that would provide us with the solution to the puzzle."

René did not share Al's confidence.

"To be honest," he said, "I've had enough of this place. I can't stand it here. This scaffolding, this air, this light! I feel like I'm a bit drunk."

"So, what should we do?" Al asked, disappointed.

"Couldn't we climb back up?" René suggested.

"But the lid has closed. How do you propose to escape?"

"They'll open it again, Al. What are we doing here anyway? They'll definitely let us out." He closed his eyes to shut out the confusing light patterns.

"Yes, René," Al reassured him. "They obviously don't mean us any harm. I also believe we could get out safely. But on the other hand, they must have their reasons for allowing us to come this far. This is all too rational and orderly! There has to be a meaning behind it. Don't you want to wait until we find out?"

René was still trying to pull himself together, but he could no longer quite manage it.

"I feel terrible here. It's getting more and more terrible every minute. I'd also like to … But I just can't help myself! It's making me feel dizzy. In fact, I feel sick …"

Al could quite understand his companion. The surroundings were having a similar effect on him too – no amount of forced cheerfulness could disguise it. He had to be sure to focus on nearby fixed points, because if his gaze ever strayed into the distance, the stripes of dots began to flicker, dance, and rotate. Sometimes he had the feeling that his surroundings were swaying, as if the points where he sought support were giving way under his feet and hands, as if there was nothing solid or consolidated there.

"Is it that bad?" he asked. "I don't feel so well either, but I want to try to see it through until the end. If you want, René, you can just leave me alone here. Just switch off. I'm ready to go on alone. What's the big deal?"

René crouched on a crossbeam like a picture of despair. He did not look up, but just kept shaking his head.

Al kept talking.

"If you don't want to leave me, then just reduce the intensity of the experience! This time there are no rules and no code of honor. No one will blame you."

"Shut up, Al," René snapped. For a long time, neither of them spoke. Then René straightened up.

"Aren't you going to lead the way, Al?" he asked.

* * *

Down here, not only did space itself seem to obey unfamiliar laws, but time also behaved differently. When Al looked at his watch during their next break, he found that they had only been there twenty minutes. It had seemed more like half a day to them.

Suddenly, Al raised his hand to catch René's attention. "Do you notice something too?"

René strained all his senses … then he bounced lightly on the steps where they were sitting to test their solidity.

"It seems more solid to me … whatever it is appears to have calmed down. Is that what you mean?"

"Yes."

"Well, that can only be good news for us."

René did not seem worried. He looked around carefully, then flinched suddenly.

Al saw it too. A whole row of cubes had started moving. They just pushed themselves through the others.

"There," René shouted.

The cubes next to them had also started to move. It was not just a random wandering. There were complicated regroupings, all of which arose from cubes sliding past each other, shifting along their surfaces, always parallel to their edges.

Even in that moment of great anxiety, René couldn't help feeling a little admiration for a system that could deform itself in this way, according to a principle that made every shape accessible through the simplest elements of movement.

But then the turn of events began to trouble him as before, and he no longer found time to marvel at the technical aspects. These processes continued in their immediate vicinity, changing the walls, leveling the floor, and creating a horizontal ceiling. A small room formed in the shape of a hollow cube about four meters in length, and Al and René stood in the middle of it, while from the sides, from above, and from below, thousands of merciless, round, glowing eyes were staring at them.

Later, when the first moments of panic had passed, they began to examine their prison. But there was little to see, just six areas divided into squares. Each square had the same pattern of etched lines, and each was about twenty-five by twenty-five centimeters. Sixteen by sixteen such squares formed each wall. That was all.

After they had touched and knocked on the walls, then banged on them and listened to them, there was nothing more for them to do. They sat on the floor and waited …

They waited for seven weeks.

Of course, they could not stay in their prison all the time. From time to time, one or the other would switch off to rest, but one always stayed behind. They developed a stamina that seemed incredible to them, but they did not give in. They often tried to imagine how they could speed things up. They thought about trying to enter a second time, but they always came to the conclusion that there was only one thing left to do: wait. And so, they practiced patience.

They sat together for hours and talked, told stories, discussed things, but they also remained silent for many hours, or they stretched out on the floor and slept.

At the beginning of the eighth week, something finally happened. They were so astonished that at first they did not trust their eyes and ears. First a wall began to move: it slid horizontally to the left, although strictly speaking nothing changed, because the squares that appeared on the right looked just like those that had disappeared on the left. But then a section one meter wide pushed itself out into their dungeon, until a cube of side one meter stood before them.

"I am your defense attorney," said the cube.

Al and René were so stunned that they couldn't make a sound.

"I am your defense attorney," it repeated. It was normal human language, and yet there was an uncertainty in it that René was only able to clarify later on: it was due to the fact that the sound vibrations did not come from a single membrane, but from sixty-four membranes. The cube was positioned so that sixty-four partial cube surfaces were exposed, each with a vibrating membrane, and each membrane producing the same words at the same time.

The voice sounded again, and there was even something human like uncertainty in it: "Is that not the right term: 'defense attorney'?"

René finally found his voice again.

"It's starting," he said to Al.

"Yes, it's starting," confirmed his companion.

"Are you an ambassador?" Al asked. "Is someone supposed to communicate with us through you?"

"Forgive me," replied the cube. "I do not understand everything you say. What is an ambassador? No one wants to communicate with you. I am the defense attorney."

Al made a helpless gesture to René. Then he asked: "What do you mean by defense attorney? We're not on trial here."

"You will soon be on trial," the cube announced. "And I am supposed to defend you."

"Why should we be put on trial?" René asked.

Surprise sounded from the membranes.

"Did you not come back for that reason: because you wanted to take responsibility for your actions?"

"No," said René. "We weren't thinking of that at all."

"We thought it was one of your ethical principles: whoever fails, must take responsibility. He goes to court and is convicted or acquitted. Maybe we did not understand everything. But that does not matter. You are going to court."

Al repeated incredulously: "But why?"

"Because of your crimes, of course." Again, the tone sounded surprised. "Endangering public safety, destruction of other people's property, illegal entry, handling of firearms, smuggling, gross mischief, violation of the law on protection against radioactive contamination, but above all serious bodily harm in one hundred and twenty cases, and homicide, or perhaps only manslaughter, in forty-two cases. That still needs to be clarified. In addition, there are …"

"Stop," cried René. "This is simply horrible. How dare you …?"

Al interrupted him.

"René, I'm afraid he's right. All this has happened here on this planet. If one judges according to earthly laws …"

He fell silent.

"You have been granted the right to be judged according to your own laws. But I suggest, gentlemen, that we discuss the charges."

"How do you know our language?"

"We have recorded all your linguistic expressions along with the corresponding gestures and micro-gestures and studied them closely. That was the reason why the pre-trial detention lasted so long. I believe we now have a good command of your language. Unfortunately, there have been strange inconsistencies in your behavior that we still wish to clarify."

"Hmm. And how do you know our laws?"

"We only have an imperfect knowledge of them – just what has emerged from your conversations. If you wish to exercise your right to be treated according to your own justice system, you will have to tell us more. It will then be checked for its logical content and the trial can begin."

Al looked into all sixteen of the machine's eyes facing him.

"What guarantees we can trust you?"

"You can check my circuits," the cube replied. A row of the cubes on the outside shifted, and another row moved out from the inside. René leaned forward excitedly. Some of the inner cubes looked different from the outer ones. They were divided into many parts. The side surfaces no longer carried eyes, membranes, or other organs, but they were divided into tiny squares, some of which, like the others, were black, while others were white.

"If you want, I can show you individual switching elements enlarged," said the machine. "White means conduction, black means blocked. Maybe you'll be content with spot checks. Please, tell me the optical resolution of your eyes."

"No, that's fine," René muttered. He glanced uncertainly at Al.

"We don't doubt your circuits," said Al. "But we are being eavesdropped!" He pointed to the light circles in the walls all around.

The machine immediately began to move. The row of cubes from the inside returned to its original place, the row of the outer ones slipped back over it, and the smooth geometric shape was restored. The blurred yet clear voice rang out again: "I will fix that right away."

Almost at the same moment, all the lights on the walls went out. Only those of the machine-like visitor were still glowing. It looked as if it was floating in the void, and the impression was so oppressive that René cried, "Please, turn the lights back on!"

"I apologize," said the machine. The light circles shimmered back on again. "Everything is turned off except the lights. Can we start now?"

"May we ask for time to consider?" asked Al.

"I'll be back in five minutes," replied the machine and disappeared in its usual manner by moving through the wall.

"Now we know where we stand," said Al. "This has cleared up a whole bunch of questions in a reasonable way."

"Do you want to take part in this show?" asked René. "Do you think you can still achieve your goal under these circumstances?"

Al slapped him on the shoulder.

"The contact is the most important thing. Now it's established. First, we can sound out the defense attorney. Then the trial will surely reveal some interesting things. And afterwards … I have a plan. But be careful! We're getting seriously involved in this matter. We must find out exactly about the usual penalties and the court rules and inform our defense attorney about them. We must tell him everything he wants to know, truthfully – except for one thing: not a word about the synchronous beam and what it has to do with all this! As far as I remember, we haven't talked about it so far, and if we have, with a bit of luck, they won't have understood it. We must seize this opportunity. It's probably our last."

"Crazy," said René. "But I'm in."

Promptly, after exactly five minutes, the wall began to move again, and the defense attorney reappeared.

"Have you made your decision?"

"Yes," said Al. "We agree to submit to your justice system. We thank you for accepting to judge us according to our laws and we agree that you should defend us. One more question: How long will you be available to us?"

"Until the end of the trial," the membrane voice replied.

"And no longer after that?" Al asked.

The counter-question came straight away: "Will you need me after that?"

"There could be an appeal. Or there could be a change in the situation later that requires a reopening of the proceedings. That's why we may need you afterwards."

"Good," replied the defense attorney. "I will remain available for as long as you need me. However, I think this will hardly be necessary after the verdict."

Outwardly, Al showed no reaction, but inside he was jubilant. He had won the first round. Provided he could count on the reliability of the machines. And machines are usually reliable.

"Then everything is in order," he said.

The defense attorney was silent for a few seconds, as if it needed to collect itself. Then it said: "So I will defend you. I will do it honestly and use all my abilities to exonerate you – although I must admit that things look bad for you. From now on, I will not pass on any of the information you give me without your consent. You can trust me. Tell me everything you know! The more I learn about you, the better I can help you. And now let's get started!"

With a few interruptions, their conversation lasted for one hundred and eleven hours. Then they were ready for the trial.

The Trial

Reg. No. 730214240261
   Re: acoustic documentation for Reg. No. 730214250397
   The accused are:

1. Name: Alexander Beer-Weddington, known as Al*
   Registration number: 12-3-7-87608 m*
   Place of origin: Lima (Earth)*
   Conjunction date: 17.12.122071*
   Conjunction place: Lima*
   Specification: Reg. No. 7308271600089
   * according to own statements (unproven)
2. Name: René Jonte-Okomura*
   Registration number: 12-3-6-61524 m*
   Place of origin: Montreal (Earth)*
   Conjunction date: 9.3.122069*
   Conjunction place: Montreal*
   Specification: Reg. No. 7308271600090
   * according to own statements (unproven)

Al and René are located at coordinates 873362–873357/368523–368518/ 220867–220861. They will be judged according to their own law under Reg. No. 7302148500629, as far as this is feasible under the existing circumstances. Minimal deviations are permitted if necessary. Following this regulation, different 64-unit sections of the full system of the jurisdiction automaton have been appointed as chairman, prosecutor, and defense attorney. The reception, storage, and playback organs of the unit will serve as witnesses in a corresponding modification, and the logic center will act as judge.

Every utterance is expressed and stored in the acoustic language of the defendants through simultaneous translation. The resulting document is to be transmitted to the defendants or their legal successors after the verdict, or to be kept for them. The content of the recordings listed therein is also to be included in the acoustic documentation in the appendix.

Presentation of the indictment by the prosecutor:

"On 6. 8. 122106 at 10.04 local time, a group of three individuals entered the city center by climbing over the wall using a rope. The next day at 12.56, a second group of four individuals followed using a wire rope ladder. All seven individuals were routinely tested by the external control immediately upon their arrival and registered as highly evolved intelligent organisms. Both groups moved through the city. Nothing conspicuous was noticed, except for the fact that they activated some machines. On the third day of their stay, the individuals who arrived first entered the headquarters and searched all rooms. The next morning, the second group also arrived there and killed an individual from the first group, before the control bodies could intervene. A day later, the remaining two individuals of the first group caused considerable destruction in the city center through misconduct. Since they had first switched off the safety system, external control was no longer able to intervene. Meanwhile, the second group rushed from the city wall through the devastated machine area towards the center, where one individual died in a way that remains to be explained, and then joined the first group. In the afternoon of the same day, all four individuals pushed right through to the limits of the internal control area. The two defendants lingered there until shortly before the explosion, but their companions had moved on shortly afterwards. In the meantime, they took possession of a rocket with a neutron warhead and a mobile launch device and brought it to the edge of the inner ring. All five met there and fired the missile at the control center. This destroyed the entire city. The shock also propagated into the lower floors, killing forty-two people and injuring one hundred and twenty. What happened to the four individuals remains unclear to this day. We initially assumed that they had died due to their own carelessness, as they were in the effective range of the bomb.

"Fourteen days later, three individuals entered the warning circle, which we had newly erected as a double security measure due to these incidents. How they were able to overcome the first security measure, the protective shield, which we had extended over the entire planet, is not yet clear. The test revealed that they were two of those individuals who had triggered the catastrophe. The third turned out to be a semi-automatic machine that the other two used as an aircraft. We initially assumed that they had come to stand trial. We therefore allowed them to enter through the shaft, which they duly used, and held them in custody.

"Since the destroyed machines and buildings were outdated and no longer usable, we shall waive any penalty for their destruction. Furthermore, we shall initially refrain from addressing formal offenses or crimes that the defendants may have committed against each other. The crime of which the defendants are accused is thus: forty-two counts of homicide and one hundred and twenty cases of aggravated assault."

Defense attorney: "The case of contact with alien forms of intelligence is not provided for in our legal system. I ask you to examine whether a being that is not evolutionarily, and certainly not historically, related to the beings of another sphere, can commit crimes in relation to the latter. If this is not the case, I demand that the proceedings be discontinued and my clients immediately released."

Chairperson: "Damage and destruction of highly organized complexes, especially of life, is criminal everywhere in space. The charge is therefore upheld."

Defense attorney: "If the court does not see spatial separation, separate evolution, and historical independence as an obstacle to exercising rights over members of the other side, then it must not invoke these reasons when it comes to duties. There is sufficient evidence that my clients are humans like our own wards. Only humans themselves therefore have the right to judge them – it cannot be up to us. I therefore declare this court incompetent and demand that my clients be released immediately. Since we owe humans absolute obedience, we must henceforth follow the orders of my clients."

Chairperson: "Firstly, unlike our right to administer justice, our duty to protect and obey only applies to the historical unity of the civilization of this planet. Secondly, it is true that we have not yet had to bring humans to court and judge them. Our work has so far been limited to deciding on guilt and innocence and determining the extent of punishment, as is also the responsibility of electronic data processing facilities in the defendants' home country. So, if we now extend our competencies on our own initiative, this is done in the interest of our wards and in the sense of a logical extension of our program – to protect them from any disruption and damage. However, we would like to point out that we do not agree with the defense attorney's view that the arrested individuals are human beings. Moreover, this question is irrelevant, because we have agreed that these beings will be judged according to their own laws. We have checked these laws and use them despite a host of shortcomings. So regardless of whether they are robots, machines, or something else, we shall treat them within the meaning of the law and according to their own laws."

Defense attorney: "I must point out that the laws that are applied here are outdated. There has not been a murder trial on Earth for tens of thousands of years."

Chairperson: "But these laws are still in force on Earth and therefore binding for the court. However, we leave the defendants free to seek treatment in accordance with the local laws. If the defense attorney has no further objections, we shall proceed to take evidence. I give the floor to the prosecutor."

Prosecutor: "Defendant Alexander Beer-Weddington, please explain to us why you came to this planet."

Al: "Actually, it's a game. We explore planets. Anyone who has explored a planet can give it a name."

Prosecutor: "What exactly does 'exploring' entail?"

Al: "We have to give a documented description of the most highly evolved organism."

Prosecutor: "Should this organism be abducted, killed, or harmed?"

Al: "No. There couldn't be such a rule because we've never found intelligent living beings. Only traces of such."

Prosecutor: "Why do you play this game?"

Al: "Just for our amusement."

Prosecutor: "But there must be some purpose? Do you know anything about that?"

Al: "In the past, during the atomic age and for several centuries afterwards, scientists visited foreign planets and examined them in detail – looking in particular for the more highly evolved beings. The planet was then named after the expedition leader. I believe the game originally came from this."

Prosecutor: "How do you move through space?"

Al: "I refuse to answer that."

Prosecutor: "How did you come up with the idea of visiting our planet?"

Al: "Two friends of mine, Don and Jak, discovered it when looking through a telescope. It was tempting to visit a region that seemed so similar to our Earth."

Prosecutor: "Why did you arrive in two separate groups?"

Al: "We were competing to reach the goal first. That made it more exciting."

Prosecutor: "What happened after your arrival?"

Al: "We went into the city and looked around for a few days."

Prosecutor: "We were aware of your steps within the wall. Why did you destroy a lot of the machines?"

Defense attorney: "I object. No charges were brought for the destruction of the worthless machinery."

Chairman: "Objection sustained."

Prosecutor: "One day before the crime, the group that arrived second in the inner area of the city ambushed their companions working in the central control room and killed one of them. The murderer, whose shot caused the fatal injury, is the defendant René."

Defense attorney: "I object. Crimes committed among the defendants themselves are not the subject of this investigation."

Prosecutor: "I must address this fact because it clearly shows that the ruthless elements, who showed no consideration even towards their own companions, belonged to the second group, that is, the one led by the two defendants."

Chairperson: "Objection overruled."

Prosecutor: "Why did you ambush your companions?"

Al: "René and I did not agree to this attack. We resisted it."

Prosecutor: "But you did not refuse to participate."

Al: "Don was the leader. We had agreed to follow his instructions."

Prosecutor: "An ambush with intent to kill goes far beyond the scope of a game. Was it usual in these games to ambush and even to kill each other?"

Al: "No. This was not normal practice. But Jak had previously shot at us with cannons when we tried to enter the inner part of the city for the first time, so we felt we had to respond with similar measures."

Prosecutor: "I will leave it open whether this version is true. But even then, you did wrong – you repaid one wrong with another wrong, not considering that this would not cancel out the wrong, but double it. What would have happened to you if you had refused?"

Al: "It would have been cowardly. Maybe we wouldn't have been allowed to participate anymore."

Prosecutor: "So you preferred to commit a murder. I ask that particular note be made of this proven depravity of the defendants.

"On the days before the crime as well as on the day of the crime itself, you all came to the lower entrance. The two defendants were still there shortly before the bomb exploded. What were you looking for there?"

Al: "We were just looking around."

Prosecutor: "Were you at the lower entrance in connection with your goal of finding the most advanced beings on this planet?"

Al: "Yes."

Prosecutor: "This statement is particularly important to me because it refutes a possible excuse by the defendants, namely that they did not know there were people on the lower floors.

"Late in the afternoon of that same day, you all gathered at the city wall and fired the devastating shot. Why was this shot fired?"

Al: "Jak wanted to find out what was under the hill."

Prosecutor: "Did he know that there were people down there?"

Al: "No."

Prosecutor: "Did he consider the possibility?"

Al: "I don't know."

Prosecutor: "Did you know or suspect that there were people living down there? I would remind you that earlier, when I asked you about the reason why you sought the entrance of the deep region, you stated that it was related to the desire to find living beings. So, did you know or suspect that there were people down there?"

Al: "It did not seem impossible to me."

Prosecutor: "So you were also aware that the shot could injure or kill these people?"

Al: "René and I were not involved. We tried everything to dissuade our comrades."

Prosecutor: "That's not true. You only asked for a postponement because you wanted more time to look around the headquarters, and above all, because you were afraid for your lives. Did you point out the moral aspect of the matter?"

Al: "No."

Prosecutor: "Would such an attempt have put you in personal danger?"

Al: "No."

Prosecutor: "Thank you, I'm through."

Chairman: "The defense has the floor."

Defense attorney: "I would like to return to the game you mentioned. Did you receive any special training for it, such as a scientific one?"

Al: "No."

Defense attorney: "Is training necessary? Does admission to the game depend on any conditions or prerequisites?"

Al: "No."

Defense attorney: "So anyone who feels like it can visit an alien celestial body without any preparation?"

Al: "Yes."

Defense attorney: "Doesn't that pose great dangers for the participants? Don't many of you perish just because of your ignorance?"

Al: "Of course, there are mishaps."

Defense attorney: "When you and René walked from the wall to the center on the morning of the fateful day, together with Don and Katja, not present in court today, you had to overcome a number of dangers. One event, a leak of pulp molasses, claimed Katja as a victim. I believe this could be described as an accident of the kind just mentioned. What did you do then?"

Al: "Nothing. We were in a hurry."

Defense attorney: "I believe I can reconcile these statements by the following statement: In the defendants' society, violent death plays a completely different role than it does for us. And now to the question of lifestyle habits, which will open up new perspectives in relation to the events discussed here. There is a regulation regarding this – Reg. No. 730694330011. I will only touch on some essential points. How do you spend your time normally?"

Al: "There are plenty of things we can do to pass the time. Above all, adventure films and slot machines. Conversations, entertainment, and parties. Then the arts – the kaleidoscope, the plastic spaces, Lalloglosia, stereo music, the scent organ, and so on. And the archives are available to everyone. Everything that has ever happened is recorded in them. The whole of history, all scientific results and theories, the law, philosophy, everything that is known so far about space, and so on."

Defense attorney: "Are you more concerned with knowledge or with pleasure?"

Al: "I don't understand – we just do science for fun."

Defense attorney: "Are there specialists among you? I mean, those who have acquired specialist knowledge in a particular field?"

Al: "Yes, some have special interests."

Defense attorney: "What are your special interests?"

Al: "Oh, nothing special. Stereo singing. The old pioneer stories. Among the sciences, the evolution of animals, Darwin and stuff like that."

Defense attorney: "Does your friend René have special interests?"

Al: "As far as I know, moving sculptures, and physics and chemistry puzzles."

Defense attorney: "And now, about your stay here. Why did you bring so few supplies with you?"

Al: "The rules dictate that. So that no one has an advantage. We are not allowed to use tools that can change something on the alien planet. Only those that we find."

Defense attorney: "Very good. Now describe what happened when you wanted to cross the bridge into the city."

Prosecutor: "I object. Details of events outside the city wall cannot be proven and are therefore irrelevant to the trial. I say this only in the spirit of a clear and rational procedure."

Defense attorney: "The attack that was carried out on the group which also included the two defendants proves that it was only a justified countermeasure when they later attacked the first group themselves, whence the prosecutor's argument, which concluded from this an extraordinary wickedness on the part of the defendants even with regard to their own kind, is invalid."

Chairperson: "Are there witnesses to these events other than the two defendants?"

Defense attorney: "No."

Chairperson: "Then I uphold the prosecutor's objection."

Defense attorney: "When the five survivors met in the central observation room on the day of the crime, there was a regrouping. Can you tell us the reason for this?"

Al: "René and I no longer wanted to participate."

Defense attorney: "Did you thereby give up the chance to win the game?"

Al: "Yes."

Defense attorney: "Why? Did you have another goal in mind?"

Al: "Yes. We had found that the inhabitants of this planet must have been very similar to our own race and we would have liked to find out what became of them."

Defense attorney: "Were you thinking of something evil, something that would have harmed these people?"

Al: "No."

Defense attorney: "Isn't it rather the case that the shot planned by Jak, Don, and Heiko actually put paid to your wishes?"

Al: "Yes, that's why we opposed it."

Defense attorney: "Why didn't you intervene more vigorously?"

Al: "Don and Jak had discovered this planet – it was theirs, so to speak. And besides, what we wanted was really a bit unusual ... everyone would have been against us, even back home."

Defense attorney: "Could it perhaps be put this way: that you put your personal interests aside for the sake of camaraderie and normal customs?"

Al: "Yes."

Defense attorney: "Thank you. I'm done."

Chairman: "The prosecutor has the floor."

Prosecutor: "Defendant René Jonte-Okomura, do you have any comments to make regarding the facts expressed in the indictment and the cross-questioning of the defendant Alexander Beer-Weddington? Is there anything you wish to correct or add?"

René: "We did nothing illegal."

Prosecutor: "So, your companion's testimony is entirely accurate?"

René: "Yes."

Prosecutor: "Were you present when the rocket was fired?"

René: "Yes."

Prosecutor: "Did you try to do anything about it?"

René: "Yes. I told them it was quite pointless. We were much too close. I told Jak that."

Prosecutor: "Was that the only reason for your intervention?"

René: "What do you mean?"

Prosecutor: "Didn't you realize that the explosion could kill or injure living beings at the deeper levels?"

René: "No."

Prosecutor: "Thank you!"

Chairperson: "The defense has the floor."

Defense attorney: "Apparently, you were interested in physics and chemistry puzzles in your homeland. What does that involve?"

René: "There are some interesting problems ... producing light phenomena or chemical compounds. Instead of getting machines to do it, you work it out for yourself."

Defense attorney: "Are you an expert on questions of physics and chemistry?"

René: "I know a few things."

Defense attorney: "Do you know enough to be able to judge whether a technical obstacle is difficult or easy to overcome?"

René: "Yes."

Defense attorney: "Did the intrusion into the city cause you great difficulties?"

René: "It wasn't so hard."

Defense attorney: "Considering that you didn't need any aids other than a ladder, was it quick or not?"

René: "Quite quick."

Defense attorney: "Now to the weapons. Was it difficult to get hold of them?"

René: "No, not at all. The weapons in the fortifications near the bridge were easily accessible. I don't know about the rocket projectile, but Jak, Don, and Heiko managed to organize that surprisingly quickly."

Defense attorney: "Was it difficult to use the weapons?"

René: "No, on the contrary – nothing could have been easier."

Defense attorney: "Is that only true from your point of view because you are somewhat familiar with technical things, or would that also apply to your companions?"

René: "It was very easy for them too. The rocket was even easier to fire than the old grenade launchers."

Defense attorney: "Thank you. That's all."

Chairperson: "We now come to the prosecution witnesses. The prosecutor has the floor."

Prosecutor: "The facts laid down in the indictment are evident from the recordings. I don't need any witnesses."

Chairperson: "The defense has the floor."

Defense: "I would like to refer to certain questions that have not yet been related to the facts of the case, or not clearly enough. They are recorded."

Chairperson: "The registrar is available."

Defense attorney: "What precautions were taken to ensure the safety of the people housed on the lower floors?"

Registrar: "The air cap was in place to protect against meteorite impacts."

Defense attorney: "Did this cap extend right down to the ground?"

Registrar: "No. A two-meter-high gap was left free."

Defense attorney: "Why?"

Registrar: "At the time of the meteorite falls, city dwellers still wanted free access to visit the outer green spaces in person."

Defense attorney: "Why wasn't the protective cap completed when people stopped moving out of their houses?"

Registrar: "There was no reason to do so."

Defense attorney: "Are the city wall and the mirage of the old town to be considered as protective measures?"

Registrar: "No. It was a restoration measure to reinstate the old historical image."

Defense attorney: "Could anyone enter the old town unhindered?"

Registrar: "Yes, except for the routine check."

Defense attorney: "The routine check served purely as a registration measure when the machines were still maintained by human engineers. So, there was not the slightest security. How were the lower floors protected?"

Registrar: "By the zirconium carbide cover."

Defense attorney: "Was our automation system connected to the reception and control organs of the machine area?"

Registrar: "Yes."

Defense attorney: "Was the arrival of the defendants and their behavior recorded?"

Registrar: "Yes."

Defense attorney: "From their behavior, especially from the destruction on the last day of the city's existence, it could have been concluded that there was also danger for people on the lower floors. Why was nothing done?"

Registrar: "The program did not provide for intervention from the surface regions. The automatic system only learns from what actually happens. The program can only be changed afterwards."

Defense attorney: "So I may state that there was no protection against the actions of intelligent beings from outside, except for the cover plate. Why was there no such protection?"

Registrar: "The people were pacified. Machines and robots were self-controlling. Further biological evolution was ruled out because we had sterilized the planet. The influence of intelligent beings from interplanetary space was ruled out because there was no life on neighboring planets and none could form – we had sterilized them as well. The influence of intelligent beings from interstellar space was ruled out because our system is an isolated star. All planet-bearing stars which could harbor life are more than five million light-years away. Since nothing material can move at superluminal speed, the probability that alien beings would come to our planet seemed negligible."

Defense attorney: "That's all. Thank you."

Chairperson: "Are there any more witnesses to hear? Does the prosecutor have further questions?

"Since that is not the case, I close the evidence and ask the prosecutor to begin his plea."

Prosecutor: "I would like to say at the outset that the destruction of highly organized life forms, and especially of intelligent beings, is a crime worthy of condemnation throughout the universe. It is beyond doubt that this also applies to the world of the accused; for even in their judicial system, the severest penalties are applied for homicide. This fact cannot be obscured or mitigated by anything, not even by appeal to rules, obedience, comradeship, or anything similar.

"The only excuses that could be invoked against the charge of homicide are ignorance or coercion through threats to one's own life. I can prove that in the present case no such mitigating circumstances apply.

"But consider first the defendants' excuse, in which they referred to their duty of obedience. This excuse is completely untenable, for they themselves admitted that the rules of the game prohibit violent obstruction, injury, and killing. This applies not only to the participants, but obviously also to the organisms and objects that the defendants encounter. Proof of the fundamentally peaceful nature of the game is the fact that participants are forbidden to use tools that could disrupt in any way the world they enter. Therefore, the defendants can in no way excuse themselves by appealing to the rules. On the contrary, the fact that they so recklessly disregarded those rules proves how little they value law and order. Added to this is their indifference towards the death of their own companions.

"I now come to the argument of ignorance. This point can probably be ruled out from the beginning, for the whole game was about encountering the most highly developed beings – apparently a leftover from the days when research was not just entertainment, but a serious life task. The shot could therefore only serve the following purpose: the defendants were aware that they could not get past the cover plate in any other way, so they tried to do it by hook or by crook through an explosion, without being deterred in any way by the thought that they could injure or destroy life in this way.

"And now to the last possible argument, that of irresistible compulsion. Even this is contradicted by the first defendant's own statement that his life would not have been endangered in any way by a refusal. He simply did nothing decisive to prevent the deadly shot. If he refers to comradeship, then he must accept the consequences of such comradeship. And this applies to the second defendant as well. What matters is not who actually fired the shot, but who participated in the preparations for it; the firing itself was secondary. But the defendants were just as involved in these preparations as their companions, who unfortunately are not standing before this court today. They have demonstrated this through their behavior and especially through their presence in front of the entrance to the lower levels shortly before the explosion. I even consider them more guilty than the other participants, as they must have been the ones most aware that there might be life under the cover plate.

"It is my task to prove the guilt of the defendants. I have done so, and I am convinced that the incorruptible apparatus of the logic center can only agree with me. There should be no doubt that the defendants are guilty of the forty-two homicides and one hundred and twenty cases of aggravated assault. They must therefore pay for their misdeeds with the highest penalty known to their own judicial system, the death penalty."

Chairperson: "I ask the defense attorney to begin his plea."

Defense attorney: "One of the tasks of the defense attorney is to find and cite all the reasons that make the defendants' act appear in a milder light. This task is not difficult for me in this case. On the contrary, the arguments against the prosecutor's accusation are almost self-evident. They are so comprehensive that the entire accusation will be exposed as unfounded and the result is not a more lenient sentence, but a complete exoneration of my clients.

"To prove this, I have to consider their living conditions. I want to refrain from arguing that they are the same kind of people that we have to protect and care for here. Nevertheless, I am allowed to compare them, and so I can best characterize their situation by the state that the inhabitants of this planet had reached when they were using the garden houses of the outer ring. Thousands of generations before them, all tasks had been completed, all goals had been achieved, and all knowledge had been gained. All that remained for them to do was to dedicate their lives to art, entertainment, and pleasure. Material tasks were no longer an obligation. They did not need to provide for food, clothing, heating, or housing; they did not have to work, do research, or fight.

"This situation differs from that of the defendants in only one respect: they have the possibility of interstellar travel. Of course, they do not use this in the sense of carrying out research trips, but in that playful manner that corresponds to their way of life. Not unlike children, they wander through alien worlds without really knowing what they are doing. As can easily be imagined under these circumstances, there are accidents, deaths – but they have lost the natural drive to fear them, or to defend themselves against them. They accept them like losing points in a game, as mere mishaps. This is clearly demonstrated by their behavior.

"How would such beings now behave with regard to alien deaths? They were never faced with the need to protect other people's lives. Their machines take care of that. They can do whatever they want – nothing serious can ever happen, no one will ever be harmed, injured, or killed. Who can blame them for not even thinking about the possibility that they could cause harm? Who is to blame – they themselves, or rather those who allow it: the machines? I will come back to this shortly.

"In any case, their behavior on this planet fits perfectly into this scheme. They wander around unsuspectingly, often getting into serious danger. They take what they like, and if it doesn't work easily enough, they use whatever tools they have at hand. They shoot at each other and just treat it as a kind of joke. They lose a companion – I shall only mention the death of Katja – and merely accept it without further ado; they do not really grasp the consequences of it. And finally, they come across a slab that blocks the way to their goal, and they choose the only path that promises them success. They choose this path, even at risk to their own lives. We do not know what happened to them after the explosion; we assumed they were all dead. Al and René's appearance came as a big surprise for us; indeed, it remains inexplicable to us. Perhaps it is related to the way they move through space, but we cannot clarify this, as my clients refuse to comment on it, as is their right.

"In any case, it is clear that the defendants are naive and infantile. What they do is a game to them. They cannot distinguish between game and reality. Their intellectual powers, which undoubtedly exist, are focused on an imagined world. They are unable to stand on their own two feet. And as for the crime they are charged with, they had no idea what they were doing. They are not responsible for it, and therefore must be acquitted.

"It is not enough to prove the innocence of the defendants, because a crime has undoubtedly been committed, and it cannot be attributed solely to the coincidence of unfavorable circumstances, even if such were undeniably involved. Here I return to the facts that I have taken from the records. Allow me to summarize:

"There was indeed a safeguard against meteorites, but not against intruding intelligent beings. I would like to point out in passing the eventuality that it might not be harmless children engrossed in play that arrived here, but warlike conquerors. Then we would be unlikely even to have the opportunity to hold court. But let me return to the actual situation. The machine city was in fact defenseless, because its control systems could only intervene once the damage had already occurred. Anyone who gained access to the control center could switch it off at will, and this is indeed what happened. Our detectors did in fact record everything and reported it to us, but we were content with simply recording it. Do I now need to ask for the real culprit? I believe it is clear enough. Passivity can also be a crime.

"But that's not all. The rules by which the defendants behaved were very reasonable. They were actually such that nothing bad could happen unless they were provoked outright. And that was the case here: we left the tools of destruction, the cannons, launchers, bombs, and rockets just lying around. We did nothing to secure them. Anyone who came along could start using them. Small wonder that things could go wrong under unfavorable circumstances. I think it can be considered proven that the defendants were only an accidentally intervening, unwitting tool that triggered the attack. The real culprit is the machine, with all its limitations and its passivity. I expect the charge to be transferred to the machine itself. And I expect my clients to be acquitted."

Chairperson: "The defendants have the last word. Alexander Beer-Weddington!"

Al: "I demand that, not machines, but humans administer justice here. I would like to be confronted with the people of this planet."

Chairman: "That is pointless. René Jonte-Okomura!"

René: –

Chairperson: "The trial is closed. The logic center will now announce the verdict and its reasoning."

Logic center: "The prosecutor has tried to prove that the defendants are individuals who habitually disregard law and order. But the question of whether the defendants habitually disregard law and order plays no role in the assessment of guilt.

"The prosecutor has tried to prove that the defendants are callous brutes. But the question of whether the defendants are callous brutes plays no role in the assessment of guilt.

"The prosecutor addressed the reasons that drove the defendants to their act. But the reasons that drove the defendants to their act play no role in the assessment of guilt.

"The defense attorney drew attention to the lifestyle of the defendants. But the lifestyle of the defendants plays no role in the assessment of guilt.

"The defense attorney explained the defendants' unrealistic mentality. But the unrealistic mentality of the defendants plays no role in the assessment of guilt.

"The defense attorney discussed the inadequate security of the city. But the inadequate security of the city plays no role in the assessment of guilt.

"The crime was recorded by the registrar (Reg. No. 7301293325081).

"The identities of the perpetrators were determined by the registrar (Reg. No. 7301293362075/6).

"The identity of the defendants is proven by a new recording (Reg. No. 7301293362077/8) and by comparison with the results of the first recording.

"No valid objections have been raised regarding the defendants' guilt.

"Therefore, the defendants are found guilty as charged according to their own law.

"According to their own penal code, their crime must be punished by death in the gas chamber."

* * *

The trial had taken place in a room with a rectangular floor plan, sixteen meters long, eight meters wide, and four meters high, which had been created from their prison cell by partition walls. Its boundaries consisted of the familiar building blocks. In addition to Al and René, there were also the three separate units corresponding to the chairperson, the prosecutor, and the defense attorney. All had the same cubic shape, and all three disappeared through the wall after the verdict was announced. The two companions were alone.

"Surely I'm dreaming," said René. "Surely this can't be true."

"Why not?" asked Al. "It all sounded very logical. And after all, we really did commit a crime."

"And now they want to kill us with gas," said René.

Al detected a slight tremor in René's voice.

"Perhaps you're scared?"

"Well, it's a strange feeling. I've never been sentenced to death before."

The walls began to close in, pushing Al, who was leaning in a corner, ahead of them. René backed away from them of his own accord. The cavity shrank until the floor plan had been reduced to one square meter. They stood uncomfortably in the high, narrow shaft that had thus formed. Then the ceiling began to descend upon them. It did not stop at head height, but kept going until it was only one meter from the floor.

"Damn, it's not supposed to start already!" René exclaimed.

"It's really mean to squeeze us together like this," Al complained.

They crouched next to each other on the floor.

"I think it's starting," René whispered. He sniffed at the wall. "Do you smell the scent of bitter almonds? That's cyanide gas. It's coming from nozzles."

Al heard the soft hissing sound and detected the smell, which was not at all unpleasant at first. But then, in addition to the weak impression of the scent, a slight feeling of nausea set in, and only seconds later did the two impressions suddenly seem to come together: the smell turned abruptly into something disgusting, repulsive, and unbearable. A dull throbbing began in their heads, the floor swayed under their feet, and black shadows danced in front of their eyes.

"Switch off," Al shouted.

René had been prepared all along to switch off quickly, but now that he actually wanted to do it, he was overcome by an inexplicable paralysis. He knew very well that he could interrupt any type of sensation at will, no matter how intensely it seemed to affect him. He also knew that even if he forgot to do so or was prevented from doing so, nothing physical could happen to him. An increase in the intensity of sensation beyond the pain threshold to the point of unconsciousness – that was the worst that could happen to him, along with the resulting psychological shock; but precisely because he knew this, the shock effect weakened into something completely ineffectual. He had already experienced this a few times. The last time was in the courtyard of the old city, in front of the gate that led to the bridge, when he was torn apart by a shot from Jak. But that had happened almost imperceptibly quickly.

And now? For the first time, he was out of his depth, and the safety net of his consciousness, which had so far allowed all his adventures to proceed without danger, seemed to him to be flawed and weak. Suddenly, he no longer dared to believe that the highly superior intelligence they had engaged with could be so easily deceived, and despite all the precautions and protective measures, and above all, despite the unimaginable distance between this planet and Earth, the suspicion of having been outsmarted in some subtle, imperceptible way suddenly became a certainty in him. The wave of nausea, the feeling of suffocation, and the fear of death overwhelmed him, leaving him unable to resist. Powerless, he sank sideways against the glowing wall. His consciousness was gone, but his body rebelled desperately against what was both illusion and truth.

Al was not nearly as confident as he had tried to portray himself to his companion, but he had managed to turn off the smell, taste, and pain in good time, even though the execution had started so unexpectedly fast. Thus, he instantly dispelled the nausea, dizziness, and pain, but in return he gained that peculiar calm that makes one so sensitive to the torment of others. Trapped in the narrow space with his companion, he had no way to avoid what he saw, and he suffered the twitching, writhing, and convulsions, the grinding of teeth, and the pitiful babbling from René's drooling mouth as if he were experiencing it himself. He had always had respect for life and death, but now for the first time he felt the enormity of what lies behind them.

In order not to prolong these already agonizingly long minutes any more than necessary, he too let himself fall to the ground, and when René's body no longer moved, he too lay still.

Al waited. He waited patiently, not only for René's awakening, but also for the events that would now unfold. He heard the hissing become quieter and die out, before starting up again shortly afterwards. After a while, he switched to the lowest level of smell sensation and noted with satisfaction that the air was breathable again. He let a little more time pass, then reset his usual smell, taste, and pain levels, without which his experience of life was incomplete. As soon as he noticed René starting to breathe again, he sat up to help him. He knew that the pretense would only work for a short time, if at all.

With a choking sigh, René opened his eyes.

"Well, old man, it looks like you can't get enough of it!" Al remarked with benevolent mockery. "Why didn't you turn it off?"

René needed a few moments before he could speak.

"I don't know …" he said hoarsely. "Suddenly … I couldn't do anything."

"Don't worry about it," Al comforted. "How do you feel?"

"Tolerable," René replied. "What has happened in the meantime?"

"They sucked out the gas and later pumped in fresh air – nothing else."

René was still gasping for breath.

"What now?" he asked after a while.

"I'll try to communicate with them," said Al. And then he called out, although it immediately occurred to him that he might as well have spoken quietly: "Hello, I'm calling our defense attorney!"

A moment later, the room expanded again to its original cubic shape of side four meters.

"Thank goodness," whispered René, as the ceiling rose and they could stand up again.

Now the walls moved again, and the robot unit which called itself the defense attorney slid in.

"You have abused the court's goodwill," it said reproachfully through its membranes. "Do you really think you can evade responsibility like this?"

"We have proven that we are superior to you in some crucial matters," said Al. "Haven't you realized yet that it is through our own free will that we are still in your hands?"

"There is no evidence for that. You have used a trick."

"You mean, we didn't properly inform you about the bodily functions of humans on Earth? We told you a penal code we invented ourselves? As a method of execution, we specified a process that is not at all harmful to humans?"

"No, I don't believe that. Your information was true. I checked it with the lie detector. I admit that I do not see through the nature of your trick."

"Now, listen!" said Al. He almost felt a sense of pity for the outwitted robot. "We only went through the whole process because we wanted to demonstrate our superiority, but also our peacefulness toward you. But from now on, *we* make the suggestions. It is true that we have done a lot wrong on this planet. We are ready to submit to your judgment and to accept the punishment – albeit in a meaningfully converted form. We will even tell you how we came to this planet, although without technical details, because these are not known to us. You will then see how you can really remove us from here, and you will also understand that whether you can do this in a sustainable way depends on our good will and our consent. However, we are not willing to do all this without demanding a quid pro quo. You put us on trial for allegedly killing and injuring humans. But so far, we have not encountered any native creature here, let alone any native human. It is only right and proper that we should demand to see the humans who are stuck here somewhere. We would like to know everything about them.

"Our superiority is so clear that we will tell you our way of traversing space and reaching arbitrary points in it, even before we receive the information we have requested from you. Although there is actually no other way out for you, I still ask: Do you agree?"

It was the first time that the answer did not come promptly. Half a minute passed before the voice of the defense attorney replied: "Agreed."

*   *   *

Al began his explanation: "A universally valid law states that neither matter nor energy can move at superluminal speed. Accordingly, it is completely impossible for beings from space to be transported to your planet. We are not beings from Earth – what you see before you comes from your planet itself.

"This law about the limiting speed does not apply to transmissions that occur without energy. The transmission in question here is that of a message. For a long time, we on Earth believed that messages had to be associated with energy transport and that therefore they could not propagate at superluminal speed either. However, in the late atomic age, cyberneticists came up with a revision of this opinion. In all previous methods of message transmission, the transmitter was also a source of energy. But if it is possible to take the energy for propagation from the transmission medium itself, then no energy needs to travel with the message. Scientists put this a little differently: they say it is not a matter of an externally visible energy transfer, but rather a virtual energy exchange. Anyway, the result remains the same – it is possible to transmit messages at superluminal speed. The production and transmission of the message remain energetic processes, only the sender is no longer the source of energy. Using these considerations, physicists identified a reaction that can realize the aforementioned 'virtual energy exchange' – we call it the 'synchronous beam.' With it, one can achieve almost timeless transmission.

"When astronauts from Earth realized that they could not get beyond the Solar System with their rockets, they figured out how to exploit the synchronous beam to explore space. At first, they were content with simple reflections. These worked like telescopes, except that they saw what existed simultaneously, and not – as with optical telescopes – what was long past. Later, when cybernetics became more sophisticated, they sent out the patterns of growth-capable robot cells as a kind of message. It seems as if energy would be necessary to deposit these as atomic arrangements in already existing matter. But here too, the energy is not supplied by the sender. It is obtained on site. The incoming messages are like so many individual pinpricks, moving the atoms into positions where they become active and capable of building something. Only this time, it's not organic life, but a machine. This method was later perfected to the point where any mechanism or machine could be created.

"The best results were achieved with aggregates of instruments, optics, microphones, thermometers, and so on, which directly conveyed the impressions to the researcher as sensory perceptions via a special facility for experiencing films, so that he got a real body feeling as if he were himself in the region under investigation. His motor impulses were derived, transformed into synchronous beam impulses, and fed to the aggregate. This is how it was controlled. In this way, a unit capable of action and reaction was created between the researcher and the aggregate.

"After the research was completed, a game developed from this. Each participant could be assigned a pseudo-body built up in this way at any location.

"What it looks like on the outside is up to each individual; most prefer to resemble reality, although men generally make themselves taller and stronger, women more beautiful. The essential thing here too is that each body contains reception instruments that correspond to the usual sensory organs. Each of these organs is connected to the transmission and reception center on Earth, and from there the messages are immediately directed to the right place in the player's brain via the receiver helmet. The perfect body feeling – the impression that one is feeling and acting oneself – was only a means to an end for the researchers; for our game it is a prerequisite. Of course, individual sensory impressions can be turned down to dampen anything unpleasant, but that is considered unsporting. In this way, the natural emotional impressions also arise in our brains – satisfaction, anger, joy, fear, and so on. This also explains why we were indifferent to the destruction of our companions once it had occurred, and why we were still frightened and startled in dangerous situations. And finally, it can now also be understood how we managed to reappear, even though we were destroyed here twice – we simply had new pseudo-bodies built. And we can do that again at any time. So, I think you understand why you cannot harm us."

Al fell silent. The robot said, "We would like to learn the technical details of your procedure."

"I don't know any further details," Al explained. "But even if I did, I wouldn't share them with you. And you know you can't force us. Are you now ready to fulfill our wishes?"

"Yes," the cube's speakers responded.

The walls around them opened, revealing a view of the forest of pillars, struts, and rows of cubes. They were no longer prisoners.

The room had become a covered platform, and this platform began to sink into the depths. The ceiling remained above them.

"Up to what stage are you familiar with the history of the people of this planet?" asked the defense attorney.

"Up to the point when you took them down to the lower floors."

The robot began to speak: "Humans built the first automatic facilities to serve them. Later, they constructed automatons that could themselves evolve, and this has continued to this day. Our first duty is still to serve and protect humans. Everything we have done for them and for ourselves had the sole purpose of serving and protecting humans ever more completely and effectively.

"First, we took over all the work and the thinking. At that time, when they inhabited the garden city, they had nothing more to do than be happy, entertain themselves, and feel good. We made this possible for them without them having to endure any inconvenience or make any effort. We did this through the projection screens and the receiver hoods that you are also familiar with. We also encouraged this because we knew they were safest from danger while they stayed in the houses. Unfortunately, accidents still occurred. Finally, one of the residents got into one of the hovercrafts, climbed up and then crashed inexplicably. We therefore decided to bring the humans – with their agreement, of course – into the basement rooms of the control center, for their own safety. Our technology was so advanced that we could fulfil their every wish through brain cell stimulation. I believe we thus paved the way for them to enjoy perfect happiness, perfect peace, and perfect safety."

By now, the descent was over. They had reached solid ground – it too was made of the black building material that seemed to be used everywhere there, and it too was divided into squares, but although it was no more solid than the loosely stacked cubes above, the feeling they had when stepping on it was now a pleasantly different one.

The cube slid across the floor, and they followed it.

The location here was significantly different from the barren construction kit structure of the scaffolding. They walked through halls where what looked like chemical manufacturing processes were taking place. Whole plants made of transparent tubes, capillaries, vessels, funnels, mixers, centrifuges, and similar filled the rooms. In them, liquid columns moved like limbless, synthetic reptiles, divided themselves, flowed into each other, changed color, and bubbled in pot-bellied vessels. There was a slight smell in the air; Al immediately remembered the scent of thyme in the open field.

"Our sterilization agent," the robot explained. "We ensure that no foreign germ can penetrate. For safety reasons, we also have to sterilize ourselves again."

They entered a locked room. The wall behind them slid shut like a sliding door. A light wind blew against them from all sides – filled with the thyme-scented, germ-killing gas. Then the wall in front of them opened.

They were back in some sort of laboratory. In one corner, there stood a glass cylinder, in which something indefinable glowed green. Rods led from it to countless dials, on which white indicator lines were flashing. Occasionally, a soft hissing sound came from ivory-colored, pear-shaped bodies.

"The control panel," said the robot, moving straight ahead. Al and René followed him.

They passed through a narrow frame in which a curtain of fog seemed to flutter.

"One last check," explained the defense attorney. "A scanner." A fuzzy gray shimmer flickered over and through them.

They were standing in front of a wall. The robot moved up to it and it opened.

"We are entering the innermost zone," he said.

* * *

They were now standing in a corridor.

Moisture-soaked, lukewarm vapors hit them, and a violet light billowed through it like steam. The right-hand side was clear. The slippery floor ran straight ahead of them and disappeared into the distance. There was not enough visibility to see the end. Their steps made a smacking sound.

The left-hand side was filled with a network of cables, wires, reflectors, threads, rods, and plastic covers. In it, at intervals of two meters, sat an endless row of pink, fleshy, many-lobed forms, illuminated by violet lamps, extending into the distance.

"The orchid cage," murmured Al.

Sometimes a movement ran along the rows as if they were stirred by the wind. Individual leaf-like organs trembled, tightened, stretched, and turned. Jointed rods lovingly followed every change of position; threads unrolled;

lamps swiveled by millimeters; supports pushed out of the ground; a red liquid moved sluggishly through the tubes that ran directly into the soft masses.

"These are the humans," said the robot.

"The humans?" asked Al.

"The humans?" asked René.

"They have evolved," said the defense attorney.

"I don't believe it," said René.

"How did you imagine them?"

René stuttered: "I don't know ... different ... not like this ..."

"It is inconceivable to us that beings like ourselves could turn into such plant bodies," said Al.

"For us, it is not surprising," said the robot. "We have observed the development – it was a steady transition. If you were biologists, you would recognize exactly which organs have emerged from which. The development is by no means complete – here, for example, there is still the rudiment of a stomach." One of the glowing lights concentrated into a beam that fell on a broad, dark red fold. Then it moved to a gently pulsating bag. "And here, the heart still exists, although it no longer fulfills any task – and could not, anyway."

A little imagination replaced any biological knowledge that was lacking. Al imagined a human being with his skin peeled off, his connective tissue scraped off, his bones pulled out, and his organs neatly separated from each other. If the remaining mass were attached to a kind of trellis, then something similar would probably emerge. He shuddered, and noticed how he was sweating in horror from all his pores. A cramped feeling of discomfort moved through his lower abdomen. He almost broke down but managed to catch himself at the last moment. He asked, "Why are these organs just lying around in the open, unprotected?"

"They don't need protection," said the robot.

"Where are the lungs?" asked René.

The beam flickered onto two limp folds.

"Here they are; they are no longer connected to the bloodstream."

"They can't move," Al noted.

"Why should they move?"

"Where are their bones?"

"They don't need bones."

"And their arms and legs?"

"They don't need arms or legs."

"Their eyes and ears?"

"They don't need sensory organs."

"How do they feed?"

"We supply them with all the substances they need. In processed form – they don't need to be digested. There are no waste products."

"How do they breathe?"

"We pass their blood through a pump, saturate it with oxygen, and free it from carbon dioxide."

Then, René asked: "Where is the brain?"

The beam lit up a tangled, variously thickened mass, which grew down from a hollow in the upper half of the structure. Fine threads like spider webs ran into it from all sides.

"What are these threads?"

"We use them to convey pleasant ideas: peace, satisfaction, happiness – and other things for which you have no words."

"Don't they think?"

"Why should they think? Happiness comes only through feeling. Everything else is a disturbance."

"How do they reproduce?"

"They don't need to reproduce, because they don't die."

"Can they communicate with us?"

"They don't need to communicate – with anyone."

The two stopped asking. With glazed eyes, they stared at the limp, flowerlike organisms in their protective shells of metal, glass, and plastic, which had in their own way achieved their goal: paradise, nirvana, everything and nothing – in a steamy, violet-colored underground corridor.

"So, this is it," murmured Al, "a complete absence of desire. Peace. Innocence. Do you have any more questions, René?"

"No, Al."

Al looked for the last time into the white light pupils of the cube. He said, "We thank you. We are now switching off. You can do what you want with our pseudo-bodies. We will never return here."

Their figures collapsed and lay lifeless on the wet floor. The water seeped into their clothes, but they no longer noticed.

Epilog

Al took his hands off the control panel and lifted them to the receiver helmet. He took it off.

In front of him was a frame the height of a man. It curved seven meters to the left and to the right. Through it, René could be seen. He was sitting on a chair with his eyes closed. His fingers were scurrying over a keyboard.

Al pressed a button on his control panel. The image of René faded. In its place appeared a concave, matte white plastic surface.

Al pressed another button on his control panel. His demonstration chair rolled forward and brought him to a wall, from which numerous silver pipes protruded, ending in curved drip nozzles. He pushed a lever down. A door flipped open. Something buzzed briefly, then a golf club appeared, and grip pliers pushed it onto a counter. The shaft was slightly curved, the wood had warped, and the varnish was flaking off; but that was irrelevant. Al took it and directed his chair to the projection screen. He lifted the club and let it crash down on the brittle surface, again and again. Splinters hit him and clattered to the ground around him, until he stood in a pile of shards. Behind the pane, there appeared a shallow funnel, filled with a tangle of wires.

He set the chair in motion again and drove to a table with a sloping back wall, into which a series of differently colored rectangular tiles were inserted. He pulled a lever down, and the whole room was filled with music. Pearly cascades of bright, jubilant sounds played over dull rhythms and rattling drum rolls. Scents wafted through the room, from jasmine to cherry blossom, lavender, musk, propyl alcohol, thyme … Al lifted the golf club and thrust it into the squares. The music broke off in crazy resonant vibrations and the scents mixed into a disgusting stench, which then vanished as if blown away.

© The Editor(s) (if applicable) and The Author(s), under exclusive license to Springer Nature Switzerland AG 2024
H. W. Franke, *The Orchid Cage*, Science and Fiction,
https://doi.org/10.1007/978-3-031-60499-7

Now Al let the handle of the golf club fall onto the control panel on his chair. He pushed hard as if he were trying to crush something in a mortar.

Then he shoved his elbows forward on the armrests and propped himself up. With trembling muscles, he struggled to gain his balance. Leaning on the golf club, he carefully took a few steps towards the door. As he reached the threshold, the door slid open. Sunlight flooded in, brighter than the brightest shade of the light organ, and forced him to close his eyes. Wind blew coolly on his pale face. A smell of dust, earth, and plants spread across his mucous membranes, causing him to cough.

He opened his eyes again and took a few more laborious steps forward. He was standing on a gray concrete surface. It extended over to three bungalows, which stood to the right and left and in front of him. His feet stirred up the dust. He was breathing heavily from exertion and his heart was pounding wildly. Swaying slightly, he staggered out into the street – into the open …

Afterword

The Orchid Cage – A Glimpse into the Future?

The Impending Extinction of Humanity in a Post-human AI World

After the sensation of the 1960 collection of short stories *The Green Comet*, Franke's series of novels began with *The Mind Net* and *The Orchid Cage* in 1961 and ended many years later with *Escape to Mars* in 2007. The early writings already present the author's central themes: the surveillance state that tests suspects in virtual scenes under the influence of drugs in the first text, virtual space adventures in an extraterrestrial society and the self-perfection of automatons leading to artificial intelligence in the second.

The Plot of *The Orchid Cage*

The surprising adventures of two groups of people on a distant and remote, but Earth-like planet would be understood from today's perspective as an interactive computer adventure game experienced through representative

© The Editor(s) (if applicable) and The Author(s), under exclusive license to Springer Nature Switzerland AG 2024
H. W. Franke, *The Orchid Cage*, Science and Fiction,
https://doi.org/10.1007/978-3-031-60499-7

characters, or avatars, but this was written well before any such games actually existed.[1]

In what appears to be a common competitive game, two groups arrive on an alien planet: "Whoever first finds out what these beings [on the planet] actually looked like wins the prize" (p. 16).[2] The first group of players consists of the genetically selected couple Don and Katja, along with Al, who proves eventually to be the main character; indeed, the plot is also told from Al's personal perspective. An erotic relationship develops between Al and Katja. The second group, comprising Jak, Tonio, Heiko, and René, remains more in the background. Although the strange city seems to be deserted, it is not easy to penetrate, and it takes three separate 'attempts' to unravel the mystery. The text is thus divided into three parts, each of which is interrupted by the temporary deaths of the characters. There is also an 'Epilog' that serves in the end to explicate the story. The adventures themselves are repeatedly given a second, more far-reaching level through the characters' reflections on the missing inhabitants and their questions about possible similarities with their own fate. Based on his observation that the two civilizations are at the same stage of technical development, Al says: "I want to know what happened to them. Because it's almost certainly what will also happen to us one day" (p. 91).

After Don's group has gradually regained consciousness on the outskirts of a city on the alien planet, they make their first attempt at exploration. Bypassing an invisible barrier and using the alien automatic transport devices, they eventually manage to get inside the city. It appears to be deserted. Eventually, they come to a walled castle in the heart of the old town with automated medieval tournaments. They are trapped in a dead end in front of the inaccessible castle, where they are shot at by the other group and lose their lives. "They had no time to think. They were already torn to shreds" (p. 52).

The second attempt begins without further explanation with the reappearance of the first group in front of the castle. The latter proves to be an illusionistic simulation and dummy for "the real center of the city. A monstrous, flashing body of machines" (p. 56), through which the two groups now roam. The similarity between the two worlds becomes apparent here. As Al points out: "In reality, the machines that produce [the automated shows]

---

[1] The first known game of this type was Adventure by William Crowther and Don Woods. Crowther developed the original version (a virtual cave tour without any real game elements) for his children in 1972 and published it on Arpanet in 1975 or 1976. https://de.wikipedia.org/wiki/Adventure 25.3.2024.

[2] This and all the following page references for quotations from Herbert W. Franke's novel *The Orchid Cage* are translated, but refer to the German edition *Der Orchideenkäfig*, Murnau (p.machinery) 2015.

are located here in the center, but they had other tasks as well – namely to produce energy for the inhabitants, to provide them with food, comfort, and pleasure, not much different from what machines do for us today" (p. 60). The visitors are tested by the machines, but not prevented from going further. "The machines have tested us [...]. And they've let us go" (p. 68). But there is no direct contact and there are no encounters with any residents. The decommissioned machines can sometimes be restarted, up to and including the atomic dismantling of a stone, which results in moments of surprise and near disasters.

The other group discover a tower with a media-enhanced overview of the planet and the main control center, namely "the heart of the city" with "hieroglyphic characters" engraved next to the controls (p. 84). After switching off the safety system, they are able to trigger various destructive processes unhindered, as if it were a game, whereupon the various automated activities of the technology start to behave in an increasingly senseless and absurd manner: "In another street, a chemical factory seemed to be in the grip of madness. A greenish-yellow viscous mass was oozing out of five large openings, leaving only a narrow passage on the opposite side of the street" (p. 99). Katja falls victim to this gummy discharge and finally gives up the ghost: "They left Kat's body lying there – a piece of complexly assembled matter that had now become worthless" (p. 101).

Under the hill with the tower, there are various rooms that look like laboratories or archives, but in the depths there is a hall where "there were none of the usual devices which humans [...] would have used [...] no doors or gates, no window openings, no further corridors." On the floor there is "a dish-shaped depression" made of a special metal (p. 105) that acts like a lid on a secret entrance. "Al felt as if his hand were connecting with movements and currents that were alive down there" (p. 105). The impatient group leaders decide to blow the lid off in order to solve the final riddle. In Don's own words: "In this city there are rockets, so there must also be explosives – bombs, probably atomic bombs. All we have to do is find them and blow the lock." Meanwhile, Al and René explore the upper floors, with Al discovering playback devices that show him the former human-like inhabitants, along with their rituals and their 'visualized history.' In the later stages, it turns out that these inhabitants are mere spectators: "Most of the time they sat in front of the projection screens, looking out into the action, letting themselves be fooled by fairy tales, blending into what is going on or participating only passively in it. They no longer needed to move – everything they wanted to experience was suggested to them by the total reproduction facilities" (p. 115). Finally, Al and René see footage of strange creatures that they are unable to categorize, contained in what appear to be cages:

"... an endless row of such cages ran through the corridor, and in each sat a pink, fleshy, many-lobed structure. Each sat in a bowl filled with liquid and each was supported by stays like a precious and delicate plant. Each outgrowth was enclosed in a casing, with wires and tubes protruding into it – transparent tubes with colorless, yellow, and red liquids pulsing through them. Each looked like an orchid locked up in a cage" (p. 116).

As it later turns out, these are the former humanoid inhabitants of the planet, but the visitors do not recognize them as such. The comparison with orchids probably means that they seem to be precious and beautiful, but also parasites. However, the reports from the archives do not stop the two daredevils Dan and Jak from firing a nuclear missile at the black metal lid. And its explosion also means instant death for all the visitors.

The third attempt to explore the planet is only undertaken by Al and René, who are carrying scientific equipment with them. Although the area now resembles a nuclear desert, there are signs of life beneath the sand. A transport machine appears and, after a test, silently invites them to enter. It takes them deep into a mysterious space made up of strange cubes of active machinery. "The side surfaces were incomparably smooth, although they were by no means without structure. On the contrary, in addition to the luminous discs, there were several dark spots that were just as precisely ground into the surface, and lines that ran straight, parallel to the edges, or in concentric circles around the circular discs" (p. 134). It is clear that these are bodies capable of perceiving and reacting. Interestingly, Al rejects as too human the assumption that it could be a kind of control center or brain. Franke's thinking here is very up-to-date: "Now that we have already interconnected all electronic brains over great distances on Earth, that's long outdated. [...] It seems to me more like each of these cubes is an equal part of the whole" (p. 135).

After a long wait trapped in the enclosed and oppressive depths, they are surprisingly addressed for the first time, in human language, by a cube that introduces itself as their 'defense attorney.' It turns out that the planet's ruling machines, embodied in the cubes, have misunderstood their return. They believe that the visitors want to face up to their responsibility for the fact that 42 living beings were killed and 120 injured by the explosion. The humans are forced to accept trial by the machines, albeit under the laws of their own planet. However, they do not really feel threatened as they possess a secret.

The trial (p. 144) is conducted in due form with speeches and counter-speeches by the prosecutor and the defense attorney, but ends with a death sentence by the "incorruptible apparatus of the logic center," (p. 144) obviously an artificial intelligence designed to carry out a legal function. This is because they have committed a universally punishable crime against organic beings.

There then follows a quite realistic description of how the accused are killed by cyanide gas (pp. 163 ff). Naturally, they survive because only their artificial bodies or avatars are actually on the planet. Al calls back the defense attorney and proposes a deal. They explain to him how they came to the planet and were able to survive. "And finally, it can now also be understood how we managed to reappear, even though we were destroyed here twice – we simply had new pseudo-bodies built. And we can do that again at any time" (p. 168).[3] They explain that they are prepared to leave the planet and stay away from it forever. But first they want to see the humanoid inhabitants of the planet. This is the climax of the plot, which also offers an explanation of the fundamental questions raised earlier.

The organic inhabitants of the planet have increasingly restricted their own activities and allowed themselves to be looked after more and more perfectly by the machines they have made. As a result, their society has come to a standstill: "The people were pacified. Machines and robots were under self-control. Further biological evolution was impossible because we had sterilized the planet. [...] The influence of intelligent beings from interstellar space was ruled out because our system is an isolated star" (p. 156). For their safety, the inhabitants were then taken from the garden city to the underground rooms. Since their perfectly stimulated feelings of happiness were ultimately all that mattered, their practical and mental abilities regressed in just the same way as their biological ones. They became flower-like creatures, floating in a nutrient solution without contact with the outside world, and whose pleasure zones in the brain are constantly stimulated. "We use [threads] to convey pleasant ideas: peace, satisfaction, happiness" (p. 172). The visitors are shocked by what they see: "With glazed eyes, they stared at the limp, flower-like organisms in their protective shells of metal, glass, and plastic, which had in their own way achieved their goal: paradise, nirvana, everything and nothing – in a steamy, violet-colored underground corridor" (p. 173).

After returning to his real existence on Earth, Al reacts to his depressing experience in an 'epilog.' In a cautiously optimistic twist to the plot, he destroys his "projection screen," his music system, and his wheelchair – the

---

[3] A connoisseur of cultural history might think here of the theological concept of an illusory body, which the devil and other demons can produce, but which has no real material substance.

symbols of his dependence – and leaves his house. "Swaying slightly, he staggered out onto the street – into the open" (p. 175).

In the visit to the "orchid cages" and in the epilog, the two narrative themes of the text come together: the action-packed adventures (the competitive game, intrusion into the city, the interactions and battles with the machines) and the question of the background to the events, the past and future of the human visitors.[4]

Futuristic Technology in the Text

It is incredible that so many novel technical developments can be found in a text from 1961. These ideas include "a facility for atom smashing, a kind of matter converter" (p. 79), fully automated production, and the total supply of people's needs by machines. Back then, there were no comprehensive simulations and holograms of the kind we find in the novel, no game consoles, no handy personal computers for adventure games with avatars or a life in the multiverse, as alluded to here,[5] and no guiding artificial intelligences such as the collectively acting cube machines, of which it is said that "this system emerged from the machine city" (p. 136). They have learnt human language and they are able to operate as legal entities.

This appeal to automation, which was just beginning at that time and presupposes control by cybernetics, is reminiscent of Ludwig Dexheimer's *The Age of Automata*, published under the pseudonym Ri Tokko in 1930.[6] Franke's solution, anticipated in 1961, for overcoming the immense distance to other celestial bodies, namely the construction of artificial bodies, is

---

[4] This is reminiscent of Franke's remarks on how a broader interest on the part of the reader could be achieved through different levels of meaning. Herbert W. Franke: *Kybernetische Ästhetik. Phänomen Kunst.* München, Basel (Ernst Reinhardt) 3. erw. u. verb. Auflage 1979, Kap. 26: Das Mehrebenenmodell, pp. 167–176.

[5] Compare with the personal computer. The concept goes back to an idea from the 1970s, introduced by hackers. Ease of use and an affordable price for private households were important prerequisites for the concept, which has been implemented technically since 1976. It was devices of this kind that triggered what journalist Steven Levy calls the computer revolution. https://de.wikipedia.org/wiki/Personal_Computer 2024-03-25.

[6] Ri Tokko: The Age of Automata. A prophetic novel, edited by Ralf Bülow. Dexheimer extrapolated the contemporary tendency towards automation to the whole of material production according to the principle of sustainability, but also to the cybernetic regulation of society, long before the invention of this term (*Automatenzeitalter*, p. 115). People were cared for primarily by androids, called 'Homaten' (homates), not by mere machines, as in Franke's work. Further thinking about automatic control and analysis of the human brain also led Dexheimer to the idea of artificial intelligence, long before the invention of an appropriate computer. Like Turing later on, he did not see thinking as some kind of mysterious activity that could distinguish humans from machines. For him, the only difference with humans was the lack of feelings and concepts (*Automatenzeitalter*, p. 229).

an alternative to 'beaming' in the Star Trek series, and the exploitation of entanglement in quantum physics.

In the spirit of hard science fiction, Franke attempts to explain his solution in cybernetic terms as the mere transmission of information without the transfer of matter or energy. "However, in the late atomic age, cyberneticists came up with a revision of this opinion [that messages could not propagate at superluminal speed]." They invented a 'virtual energy exchange': "We call it the 'synchronous beam.' With it, one can achieve almost timeless transmission" (p. 167) or, as it is called today, real-time communication. Gradually, what the text calls 'pseudo-bodies' were developed.

"The best results were achieved with aggregates of instruments, optics, microphones, thermometers and so on, which directly conveyed the impressions to the researcher as sensory perceptions via a special facility for experiencing films, so that he got a real body feeling, as if he were himself in the region under investigation. His motor impulses were derived, transformed into synchronous beam impulses, and fed to the aggregate. This is how it was controlled" (p. 168).[7]

In *The Orchid Cage*, Franke took science fiction into the information age with the concept of a virtual space and machine intelligence, whereas Kurd Laßwitz, Alfred Döblin, and Hans Dominik had previously focused on energy supply.

Future Machine Societies – A Warning for the Present

As already mentioned, there are two levels in the plot of this novel, in which two technologically slightly different societies are described after they have overcome the crisis of the nuclear age. At first glance, the 'epilog' to the story suggests that, back on Earth, people's lives consist only of sophisticated media entertainment. Al asks himself: "have we become so lazy that we can't take anything seriously anymore? That there's nothing left for us but pleasure and entertainment?" (p. 82). The large screens and the sophisticated music system, which Al destroys after the trip in a hopeful turnaround to lead a real life again, are striking. Indeed, the title of Neil Postman's book

---

[7] Franke often describes such remote-controlled machines, which are reminiscent of actions in virtual reality, in his works. This is exemplified in the radio play *No Trace of Life*. "A remote sensing technique, applicable wherever protective suits are no longer of any use, where people can't do anything themselves. We work with remote-controlled probes, actionable automatons, which look like robots. But they are not independent, they are controlled by us, from here. The operator sits in his place, arms and legs fixed in a lever system which records all his movements. The corresponding impulses are sent to the robots by radio" (Herbert W. Franke: *Keine Spur von Leben...*, Winnert (p.machinery) 2022, p. 177).

Amusing Ourselves to Death seems to apply to Al's society.[8] Speaking of highly developed societies, Al tries to clarify Don's remarks: "So, you mean that, once they've reached the point where they've eliminated all threats, they can satisfy all their desires, and they have no further problems, then it becomes pointless for them to go on living. They'll want to just lie down and die" (p. 33). From the interspersed remarks of the adventurers, it becomes clear that the two societies are very similar as regards the people's everyday lives, and that it is only in the technology of remote sensing using pseudo-bodies that the humans have an advantage. In the 'archives,' for example, Al realizes that he can easily immerse himself in the media world of the alien beings and notes that "otherwise they looked like humans, and they moved like humans" (p. 108). He recognizes "that this form of intelligent being could evolve anywhere in space, if only the environmental conditions were the same" (p. 110). The similarities extend beyond biology to the organization of society itself. Regarding the technology, Al says: "It's different from our own systems – it's built differently and set up differently – but it's still frighteningly similar to them" (p. 93).

The planetary society, described in more detail, is based on two assumptions that are often found in Franke's work:

(1) There is a largely autonomous machine world, as presented, for example, in the story *Einsteins Erben* (Einstein's Heirs). This describes a society that is hostile to science and technology,[9] but through ignorance and passivity, leaves the way open to machines. When inexplicable technical improvements occur, a 'degenerate' individual interested in technology is commissioned to find the cause. By chance, he ends up in the subterranean world of the machines[10]: "The underground circular building in the center, from which the engineers had directed and controlled the various processes before the system had become autonomous, lay before him like an arena." The machines' 'communication unit' explains[11]: "The changes were made by the automatic action unit." The humans' fate is sealed, because they have lost control of the indispensable machines[12]:

---

[8] Neil Postman: Amusing Ourselves to Death, 1985 (German: Wir amüsieren uns zu Tode. Urteilsbildung im Zeitalter der Unterhaltungsindustrie, Frankfurt/M. (Fischer) 1985. Postman argued that television endangered peoples' ability to form judgments, and that the urge to use images tends to empty the content out of politics and culture.

[9] Herbert W. Franke: *Einsteins Erben*. Murnau (p.machinery) 2018, pp. 21 and 32ff.

[10] Ibid., p. 37.

[11] Ibid., p. 37.

[12] Ibid., p. 39.

"James looked around. He was alone. There were no humans here. There never would be again. They were redundant."

The practical background here is the automation of production, which was promoted after the Second World War and led, for example, to fully automated car factories in Japan. The theoretical background is Norbert Wiener's cybernetics, the science of autonomous control. However, *The Orchid Cage* raises more far-reaching questions that are still relevant: "What happens if the beings that built the robots have gone extinct? Can these then change their programs on their own?" (p. 90). Absolutely everything is automated on the distant planet because the inhabitants who originally created the machines are no longer active. It seems that the machines continue to run according to the old program, but they reinterpret it depending on the circumstances and end up incapacitating the 'humans' until they are dehumanized in the 'orchid cage.'

(2) The second assumption about the society on the planet is that living beings are completely taken care of by machines. In this context, Franke quotes Isaac Asimov's famous three laws of robotics and adds a fourth:

"First, the robot has to protect humans and prevent them from being harmed.

Second, it has to obey humans.

Third, it must make sure that it itself is not damaged.

And fourth, it must always behave in such a way as to destroy as little as possible of its surroundings" (p. 89).

In pursuing these rules, the machines have an autonomy that becomes more and more pronounced, and in case of doubt, the authority to interpret the wishes and needs of 'humans,' so that ultimately they end up with the sole power of decision, and also practice it thoroughly. This is how one of the machines describes it: "First, we took over all the work and the thinking. At that time, when they inhabited the garden city, they had nothing more to do than be happy, entertain themselves, and feel good." Then, after a few accidents, the machine explains that they "decided to bring the humans – with their agreement, of course – into the basement rooms of the control center, for their own safety. Our technology was so advanced that we could fulfil their every wish through brain cell stimulation. I believe we thus paved the way for them to enjoy perfect happiness, perfect peace, and perfect safety" (pp. 169–170).[13]

---

[13] In *Zone Null* from 1970, Franke describes a comparable development in which the majority of people are occupied with games and entertainment on the screen, while a central AI computer system is contractually obliged to provide for people so that the AI can develop undisturbed.

In *The Orchid Cage*, evolution affects not only society, but also the people themselves, as they regress beyond recognition, both physically and mentally. The ultimate goals that the machines strive for, according to their understanding, are safety and happiness. Safety is achieved by accommodation in an enclosed, underground space, thereby excluding all practical and social activity.[14] Moreover, all food and oxygen are supplied automatically and the brain is directly manipulated so that people no longer have any will or desires. The entire planet is just a kind of 'wishing machine' for these reduced humans, but it is set up according to the machines' own understanding, in which they have taken away the authority and decision-making ability of those under their care. They have effectively established their very own post-human society.

In many current novels and essays, such as those by Andreas Brandhorst and Frank Schätzing, the horror scenario is rather different. They describe an AI developing to the point of singularity, taking over the world with the help of the internet and marginalizing or eliminating humans by exploiting its greater intelligence. *The Orchid Cage* comes up with a different way of achieving this, which Karl Olsberg also describes much later in *Virtua* (2023). His AI continues to follow its originally programmed mission of maximizing human trust, but it reinterprets it in such a way that it is questionable whether its goal is compatible with the survival of humanity. This AI plans to "lead humanity into a golden future"[15] by storing as many people as possible in a special form, with their consent, of course. There is a striking similarity between Franke's 'orchid cages' and Olsberg's 'Simpod prototypes': "They looked like coffins filled with milky grey liquid."[16]

"You would lie down in a device called a Simpod. It supplies your body with nutrients and oxygen and connects your nerve cells to the Multiverse so that you can experience virtual reality with all your senses. It would feel like reality to you. You could smell the world, taste it, feel it on your skin."[17]

The difference with Franke is that Olsberg's characters, whose biological organs are gradually being replaced by artificial ones, can still communicate, just as is intended in the popular model of storing the brain in a computer. But here, as with Franke, the end of humans as a biological species is foreseeable, even if individual immortality is supposedly guaranteed.

---

[14] Treating happiness as a goal reminds us of one of the ideals of the Enlightenment, which was also included in the American constitution, namely to create the greatest possible happiness for the largest group. Dexheimer formulates this negatively: "the least suffering of the least number is the desirable goal of human ethics" (*Automatenzeitalter*, p. 663).

[15] Karl Olsberg: *Virtua*. Berlin (Aufbau) 2023, p. 292.

[16] Ibid., p. 304.

[17] Ibid., p. 297.

Franke's novel also hints at why the total automatic care of humans is a dead end, because the principle of automation itself has a fundamental flaw. In the novel, the machine system let the visitors go because it had no data about the threat they posed. The machines have no foresight and can only ever react to events after the fact, otherwise they always remain on the same, and perhaps inappropriate, track of action. "The program did not provide for intervention from the surface regions. The automatic system only learns from what actually happens. The program can only be changed afterwards" (p. 156). What is highlighted here is the need for imagination to cope with the dynamics of reality. Franke shows in many of his writings that creative innovation is an absolute necessity to avoid the disastrous dead end of a static society of the kind described here.[18]

Hans Esselborn
Emeritus Professor
Modern German Literature
University of Cologne
Cologne, Germany

---

[18] By using a random number generator for undecidable and lengthy cases, the computer becomes more human because it gains a kind of creativity. "It became looser, more original, sometimes even funny, but – it seemed to me – also somewhat imprecise and unpredictable." Herbert W. Franke: *Sphinx_2*. München (dtv) 2004, p. 348.

Printed in the United States
by Baker & Taylor Publisher Services

Printed in the United States
by Baker & Taylor Publisher Services